Cosmic Entity

Cosmic Entity

✦

A Timeless Perception of the Universe

Mark A. Strain

iUniverse, Inc.
New York Lincoln Shanghai

Cosmic Entity
A Timeless Perception of the Universe

iUniverse, Inc.

For information address:
iUniverse, Inc.
2021 Pine Lake Road, Suite 100
Lincoln, NE 68512
www.iuniverse.com

ISBN: 0-595-30125-8 (pbk)
ISBN: 0-595-66134-3 (cloth)

Printed in the United States of America

"There are more things in heaven and earth…
Than are dreamt of in your philosophy…"
—Shakespeare
Hamlet, *Act I, Scene V*

Contents

PREFACE

When I was a small boy I used to go out into the backyard and look up into the night sky and wonder about those little points of light. I used to think the stars were relatively close, about the size of a tennis ball and composed of some sort of bright, silvery metallic substance with sharp, thin spikes extending from the fist-sized ball. I reasoned that these tiny stellar "tennis ball" objects must be close if they are that small. Maybe if I jumped high enough I could touch one of them. I must be careful, though, because those bright spikes emanating from their surface may be sharp. I never seemed to be able to jump high enough.

Astronomers, for ten thousand years, have had their own ideas about the heavens—about their origins and about the origins of their home planet, the moon, the sun and the stars. Their ideas and mythologies seem ridiculous and childish, almost as preposterous as my early misconceptions. Cosmology is almost entirely a visual science. You cannot reach out and touch a star (luckily). A cosmologist cannot travel by any means to the nearest star or galaxy to study it. We cannot send out space probes to scoop up stellar material and bring it back for examination. The distance and time involved for such ventures are just too great. Therefore, the only tool cosmologists have to study the universe is starlight.

Light (and other forms of electromagnetic radiation, like radio waves) emanating from stars and galaxies is captured and split into its individual components (frequencies) and analyzed. A lot of information about a star's constituents can be determined this way. Again, cosmology is a visual science. This makes it a difficult subject and usually out of reach for the general audience, but a lot of progress has been made over the last one hundred years and, more recently, over the past decade thanks to Mr. Hubble's space telescope.

Starlight does not instantaneously travel from point to point. The speed of light, although tremendously fast, is finite. Peering out into space is equivalent to looking back in time. Even sunlight is old. The sun is 93 million miles away (or eight light-minutes)—just a short jaunt in cosmic terms. Sunlight has been traveling for eight minutes to reach us here on Spaceship Earth. Light from the nearest star (other than the sun) travels over four years before reaching the earth. Space, time, distance and age are all subjects for discussion in cosmology.

Cosmology is the study of the universe on a grand scale, from the extremely minute subatomic particles and their constituents to the stupendously large supercluster walls of galaxies extending across a large portion of the universe. Cosmology encompasses the disciplines of astronomy, astrophysics and particle physics, among others. The objective of this author is to attempt to bring this subject to the general audience.

The title of this work, *Cosmic Entity*, was chosen to paint a picture of the origin of the universe. The universe began from a point, an "entity" of zero dimensions, before space and time existed. From this empty entity, the universe expanded into its present "cosmic" proportions.

This author owes a debt of gratitude to all those authors listed in the reference section of this book and countless others not listed. Cosmology, as well as all other disciplines of science, is built on the foundation of research and intuition of others. It is a cumulative science. This author sincerely apologizes for any accidental errors, discrepancies or inaccuracies. The study of the universe and its origins is a difficult and complex endeavor, and not all theories or data are perfect. But the quest for knowledge, and the continual strive to improve ourselves and our level of awareness persists in the human spirit.

FROM ANCIENT MYTH TO MODERN THEORY

It is perhaps the longest sought-after question: the query of our origin. Mankind has pondered the origin of the universe since he has been able to look beyond the ancient campfires and up into the heavens. To ancient man, the night sky was a mythical place, a spiritual realm. The stars and planets were gods to be respected and feared—beings to be worshiped. Finally, beings on this planet have developed the unique ability to ponder their own existence.

Tens of thousands of years later, in the fourth century B.C., the Greek philosopher Aristotle conceived of an earth that was a round sphere. He noticed that ships sailing in from a distance from over the horizon became visible masts first. Distant sailing ships never presented their hull first, always their mighty sails. This, Aristotle reasoned, was due to the curvature of the land (and sea). Aristotle also noticed that during a lunar eclipse the earth cast a round shadow upon the moon. The only shape that would always cast a round shadow from any angle, Aristotle reasoned, was a sphere. Therefore, the earth must be round. No longer was mankind doomed to survive on a flat expanse of *terra firma*, but proof was hard to come by, and reason was insufficient for the stubborn.

Aristotle was even able to estimate the circumference of the earth by noting changes in the position of the North Star with corresponding changes in latitude. Although he was in error by almost twice the currently accepted value, his ability to deduce scientific knowledge by reason was amazing. To Aristotle the sun, moon, planets and stars circled the earth. He took the geocentric view that the earth was the center of the universe.

Claudius Ptolemy, an Egyptian living in Alexandria in the second century A.D., refined Aristotle's picture of the heavens with the cosmological model of concentric spheres. The Ptolemaic theory described the earth as an inert, immovable mass, located at the center of the universe. In this model, the earth was the center of eight concentric spheres. The sphere closest to the earth was one that contained the moon. The next sphere provided the sun, and the next spheres held Mercury, Venus, Mars, Jupiter and Saturn. The outermost shell carried the fixed stars. Each sphere moved independently, which accounted for the independent motions of the planets against the background of stars. Ptolemy's music of the spheres provided a reasonably accurate model for predicting the positions of astronomical bodies. This geocentric view of the heavens pleased the Christian Church. Humans painted themselves perfectly pertinent at the center of creation. This picture was accepted for fourteen hundred years. Enter Copernicus.

The Polish priest, Nicholas Copernicus, from the sixteenth century A.D. is considered by many to be the father of modern astronomy. He described a universe in which the sun is stationary and the earth and planets move in circular orbits around the sun. Copernicus theorized that the earth rotated once daily upon its axis and revolved once yearly around the sun. He assumed the orbits of the planets (including earth's) were circular. For this reason, he could not explain the eccentric orbit of Mars. Johannes Kepler later refined the Copernican theory by explaining that the planetary orbits were elliptical instead of circular and introduced orbital equations that precisely predicted the positions of the planets.

The heliocentric (sun-centered) view of the universe was not widely accepted at first. This was an extremely radical idea of the time, an idea that egotistical man was unwilling to accept. Up until then, mankind was at the center of the universe, set aside from nature, and few dared to challenge the medieval Church. The heliocentric theory of Copernicus led man to realize that he is actually part of nature and not superior to her. This idea struck a deep blow at the religious and spiritual beliefs of the time. Mankind was no longer at the center of the universe. The Church felt this line of reason was heresy, and was very slow in accepting the idea of man's modest existence.

About a hundred years later, the Italian astronomer Galileo Galilei supported Copernicus in the heliocentric view of the heavens. With their radical beliefs and controversial ideals, Copernicus and Galileo served to usher in the Scientific Age.

With a telescope, crude by today's standards, Galileo observed the four innermost moons of Jupiter: Europa, Io, Callisto and Ganymede. By observing the Jovian system night after night and noting the positions of the different moons, it became obvious to Galileo that these tiny specks in his looking glass were gently circling Jupiter. With a bit of insight he reasoned that the earth's moon was in orbit around the earth, and that Jupiter, as well as earth, was in orbit around the sun.

With the aid of his telescope Galileo discovered the phases of Mercury and Venus. The phases are similar to the monthly phases of the moon—crescent to full. The discovery of the phases of Mercury and Venus farther reinforced the notion that these planets orbited the sun.

This idea knocked over the central pillar to the structure of Christian religion. The Church threatened Galileo with torture if he persisted in teaching that the earth was not at the center of creation.

About the same time, the Italian scientist, Giordano Bruno, declared that space was boundless and that the sun and planets were not alone in the cosmos, but were one of countless systems in the heavens. He went on to say that on these distant worlds there could be rational beings even superior to man. For this blasphemy, Bruno was burned at the stake, and Galileo was forced to renounce his belief in the Copernican theory and was sentenced to house arrest for the remainder of his life for heresy. Only much later, following unmistakable evidence, the Church was able to accept that our round earth orbits an average star on the outskirts of an ordinary galaxy somewhere in this vast universe.

Almost four hundred years later, the *Voyager 2* interplanetary spacecraft, sent out to explore the outer solar system, has taken one last look back at her home—at our home. She saw a world, a system in which the sun is centered, as Galileo and Copernicus predicted. She saw a distant group of ordinary planets. She noticed the third planet (from which she was sent) was of below-average size. She looked back at the sun—an average middle-aged, yellow, main-sequence star on the outskirts of a common spiral galaxy belonging to a local group of galaxies, and on up to a parent cluster and giant supercluster of neighboring galaxies. What *Voyager 2* found was not a world in which we are the center of the universe. She saw earth as an insignificant part of a grandiose whole. Removing the earth from the center of creation does not dismiss God, but instead opens our eyes and minds to a vast realm of possibilities.

In 1687 Sir Isaac Newton formulated the universal law of gravitation. Newton's law of gravity, along with the laws of motion, provides the basis for a mathematical model for the motion of the planets. Newton explained how, if a cannonball was shot into the air, it would fall back to the ground due to some unseen force (gravity) pulling it back down. He went on to say that if the cannonball was shot into the air with greater force (and hence with greater velocity) it would go even farther downrange. Along the same line of reasoning, if the cannonball was shot with an incredible amount of force (and hence with a high enough velocity) it would not fall back to the ground; it would, instead, enter into orbit around the earth. If the cannonball was shot straight up, instead of at an angle, it would fall back to the earth unless it attained a certain velocity. This velocity barrier is called the "escape velocity," and, for the earth, is equal to about 25,000 miles per hour.

So, the *Saturn V* booster that carried Neil Armstrong, Buzz Aldrin and Michael Collins aboard the *Apollo 11* spacecraft had to pass the 25,000-mph barrier in order to escape the earth's gravitational influence in order to enter into a lunar trajectory. Any speed less than this would have been insufficient, and the spacecraft would have either fallen back to earth or at least fallen into orbit around her.

Albert Einstein improved on the glamorous theories of Newton. In 1905, at the age of 26, Einstein formulated his special theory of relativity. He observed that all motion is relative. If a body is in motion (in space where there are no reference points) next to another body, it is impossible to an unbiased observer to say which body is in motion. One can say only that one body is in motion *relative* to the other. Time is also relative: as an object's speed increases, its time (relative to a stationary observer) slows down. Einstein went on to say that the speed of light is a special number in nature. The speed of light is constant and independent of the motion of its source. No influence can make it go faster or slower. Einstein observed that nothing can exceed the speed of light—it is a cosmic speed limit. As the speed of an object increases, its mass increases. As the object's speed approaches (but never equals) the speed of light, its mass increases exponentially. Therefore, it would take infinite energy to increase the speed further. Einstein also theorized that the length of an object, in the direction of motion, shortens as the object's velocity increases.

In 1915 Einstein postulated his general theory of relativity. Einstein formulated an advanced new concept of gravitation. Einstein went on to say that space is curved, and that gravitational fields bend (or warp) the fabric of space. Relativity unified space and time and described the influence of matter on spacetime. With the general theory of relativity, the curvature of space and gravitational effects are understood like they have never been before.

Edwin Hubble, for whom the space telescope is named, explained that there are galaxies beyond the Milky Way. Our galaxy spirals a wonderful waltz among countless other stellar islands. Hubble noticed how galaxies are distributed in distance—the more distant, the greater their redshift. He described galaxies as the basic unit of distribution in this unbelievably vast cosmos. In 1929, Hubble attributed the redshift of distant galaxies to the expansion of the universe—an inconvenient mathematical incongruity that Einstein was reluctant to admit.

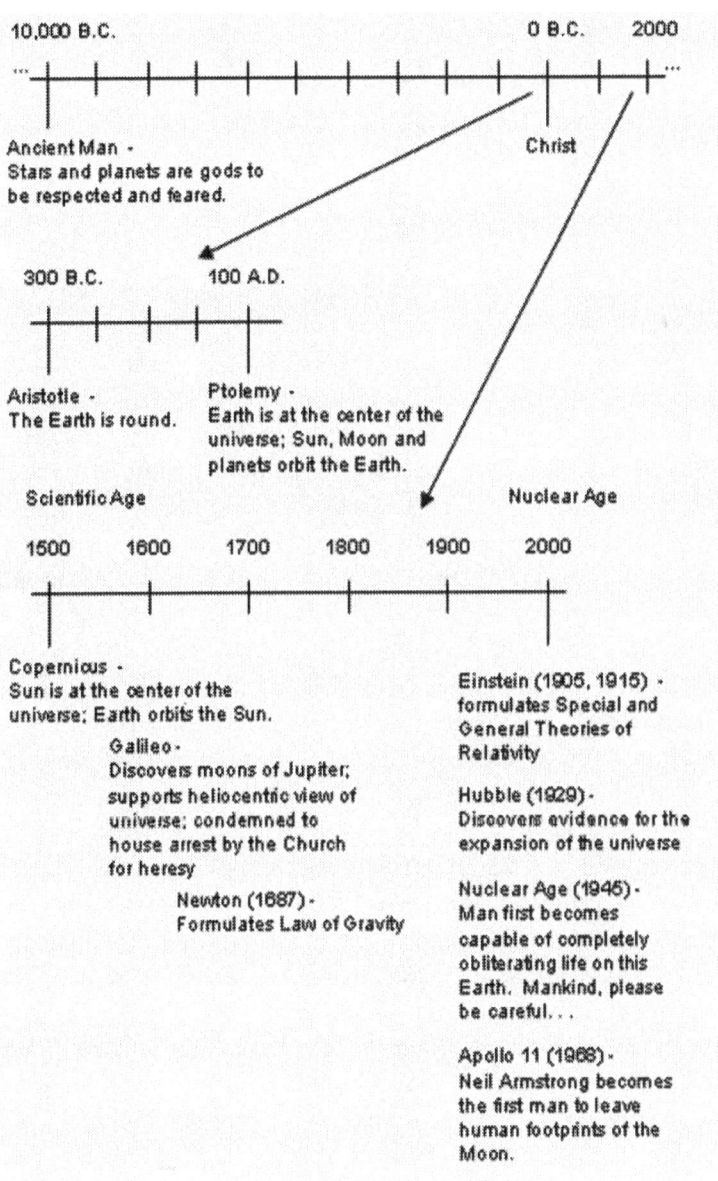

10,000 B.C.   0 B.C.   2000

Ancient Man -
Stars and planets are gods to
be respected and feared.

Christ

300 B.C.   100 A.D.

Aristotle -   Ptolemy -
The Earth is round. Earth is at the center of the
                     universe; Sun, Moon and
                     planets orbit the Earth.

Scientific Age   Nuclear Age

1500   1600   1700   1800   1900   2000

Copernicus -
Sun is at the center of the   Einstein (1905, 1915) -
universe; Earth orbits the Sun.   formulates Special and
                                         General Theories of
        Galileo -   Relativity
        Discovers moons of Jupiter;
        supports heliocentric view of   Hubble (1929) -
        universe; condemned to   Discovers evidence for the
        house arrest by the Church   expansion of the universe
        for heresy
                                          Nuclear Age (1945) -
                Newton (1687) -   Man first becomes
                Formulates Law of Gravity capable of completely
                                          obliterating life on this
                                          Earth. Mankind, please
                                          be careful. . .

                                          Apollo 11 (1969) -
                                          Neil Armstrong becomes
                                          the first man to leave
                                          human footprints of the
                                          Moon.

Timeline.

THE DARK SKY

The night sky is dark. This is such an obvious and unquestionable fact that it is almost humorous to even mention. Practically all of us take it for granted. Of course the night sky is dark. The sun has set and is no longer in the sky above us. This observation is contrary to previous misconceptions of the universe.

Up until the early part of the twentieth century astronomers and physicists pictured the universe as a static, unchanging, infinite expanse of space that has existed forever. This idea is not unreasonable, for change and large-scale motion is not apparent, and astronomy is an observational science. A universe that has existed forever, containing stars equally as ancient, must lead to certain ridiculous conclusions.

The German philosopher Heinrich Olbers, in the early nineteenth century, realized the discrepancy that the night sky is dark. This incongruity is known as Olbers' Paradox. An infinite universe that is uniform must contain an infinite number of stars. A universe consisting of an infinite number of stars should be permanently bright, so the entire night sky would be as bright as the sun. Why, then, are the heavens dark?

If the universe contained an infinite number of stars, the sky would be forever bright because every direction one looked in the sky would eventually end up at a star or cloud of hot interstellar gas. What about the heat? An infinite universe with infinite stars radiating forever should have heated up the entire universe to a temperature as hot as the sun. This is the paradox.

In actuality the universe is a dark and frigid place, on average. The average temperature is just a few degrees above absolute zero. The misconception is that of an infinite universe. The initial assumption of a static, infinite universe must be flawed. The universe must be finite in size and in time (it must have had a beginning). The universe cannot have existed forever because the light from distant stars has not had sufficient time to reach us since the speed of light is finite. This is why the night sky does not shine as bright as the day. Stars have not been radiating forever, and light from distant stars (in distant galaxies) has not had enough time to traverse the great distance. Stars have a limited lifetime. They are born, shine for hundreds or thousands of millions of years and they end their lives—some with a whimper and others with a bang. Therefore, they could not have heated up the universe to millions of degrees because relative to the age of the universe they are short-lived.

Another discrepancy plaguing Olbers' Paradox is the idea of an unchanging universe. A finite, static universe would be overcome by the overwhelming urge to collapse. A finite universe must be expanding or else galaxies would be moving closer together by gravitational attraction. The universe would have collapsed long ago. Olbers reasoned that the universe must be finite in size and age because it is dark, and expanding because it has yet to collapse. These ideas today are commonplace. The universe is finite and expanding, which implies that it had a beginning.

BEGINNING TO END

What existed before the universe? Cosmology abounds with questions. It is difficult and maybe even impossible to say for sure. It is difficult to know what existed before the creation of the universe. The best description of this reality is nothing—no space, no time. Perhaps some other universe in some other plane of existence ended its life in an endless cycle of destruction and creation leaving the potential for the Big Bang. Maybe when a universe with enough mass reaches a certain size and diminishing momentum it begins to contract smaller and smaller (like a Big Bang in reverse). If this universe reached a point of infinite density and energy—an infinitesimal point—maybe the situation would reverse itself, initiating a Big Bang. On the other hand, perhaps the universe that exists today is the only one that has ever existed and the only one that will ever exist. It is extremely difficult to say. We may never know.

Space and time, which constitute the spacetime fabric of the universe, were created at the Big Bang—the genesis event that sparked creation. Before this, space and time had no meaning. Undeniable evidence exists today to prove this epiphany—the expansion of the universe, the existence of the cosmic background radiation and the abundance and even distribution of the element helium throughout the universe.

Unquestionable proof exists for the expanding universe. The fact that it is expanding implies that the universe had a beginning sometime in the distant past, and it suggests that the universe is finite. Something had to set this sequence of events in motion. A static universe (an unchanging one that neither expands nor contracts) gives no evidence of earlier times, no reason for a beginning. Our expanding universe had a clear and definite beginning somewhere between ten and fifteen billion years ago. A universe that is expanding (or any other expanding volume for that matter) had to begin from somewhere, from something.

We live in a day and age when (in most parts of the world) scientific discovery and beliefs will not get you sentenced to house arrest (like Galileo), or worse yet, burned at the stake (like Giordano Bruno). Cosmology is the study of Mother Nature on a grand scale. Applying science to unravel some of nature's secrets does not remove any amount of religion from her. It is often convenient to use religion as a reason for not comprehending the workings of the universe. Understanding bits and pieces of the universe to picture the bigger whole does not remove any piece of religion. It only serves to respect what a wondrously ancient and grand entity the universe really is. Perhaps it is God's will not for us to remain ignorant and unknowing, but to study, discover and accept what our eyes and minds tell us—that the universe is a majestic gift of space and time.

Georges Lemaitre—engineer, cosmologist and Roman Catholic priest (ordained in 1923)—saw no conflict between faith and modern science. He believed that one needs to realize that the authors of the Bible did not intend for it to be a textbook of science, and that once this notion is accepted, the controversy between religion and science vanishes. We should not abandon religion or science because we misinterpret what the author of Genesis, Chapter 1 interprets as the definition of a day.

The previous theory of the universe, the Steady State Theory, describes the universe as infinitely large and infinitely old. According to this theory, matter is continuously created in voids between galaxies to maintain a constant density as the universe expands. This theory has been practically rejected by modern cosmology. An infinitely old universe also contradicts the teachings of the Church that states that the universe ("the heavens and the earth") had a clear and definite beginning. Modern theory also dictates that the universe is finite in age. Here science and religion agree.

Big Bang cosmology describes a universe created in an instant from nothing. Theology seldom discusses the raw material that God used when engineering creation, so this point is not in dispute either. The term *ex nihilo* (out of nothing) describes the creation of the universe from nothing—no energy and no raw material. This old adage has been used against religion to say "without God," but the notion that what we think of as the universe could be created from nothing points to God. Here science and theology intersect once again. Cosmologists now believe that the universe was created from something called a "false vacuum" where the sum of matter and energy equates to zero—nothing. Matter is considered "positive energy" since matter and energy are different manifestations of the same entity, and gravity is considered "negative energy." The sum of all positive energy and negative energy in the universe equates to zero. In a nutshell, the universe originated from a point over ten billion years ago.

Cosmology is the study of the physics of Mother Nature, from the infinitesimally small to the stupendously large, from the miniscule structure of matter itself to the colossal large-scale structure of the universe. Modern cosmology describes the universe by the cosmological principle—that the universe is evenly distributed and looks the same in all directions. The structure of the universe was determined in the first few seconds after the Big Bang. Miniscule fluctuations in the density of the matter field expanded with the expansion of the universe and became the seeds of the colossal structures evident today—superclusters of galaxies.

Hydrogen was created shortly after the Big Bang. Protons are not synthesized in stars. All matter (protons) that exists in the universe today came from the primordial soup left over shortly after the Big Bang. This hydrogen served as the building blocks of all matter evident in the universe today—from red supergiant stars to the moons of Jupiter and little puppy dogs. The early universe was dominated by photons during the nucleosynthesis phase. The tremendous heat and pressure of the infant universe fused a great deal of hydrogen (protons) present to form deuterium and helium in the first few minutes of the universe.

The universe would have to wait another billion years or so for heavier elements to be synthesized in stellar cores. Galaxies soon formed to harbor these stars, and these "island universes" became the basic unit of structure in the large-scale universe. Galaxies became the building blocks of larger structures: clusters of galaxies, superclusters and supercluster complexes that span a large portion of the visible universe today.

Questions facing cosmology today focus on the expanding universe, the actual age of the universe and the amount of matter contained within the universe. What will be the fate of the expanding universe? Will it halt? How ancient is the universe?

SPACE AND TIME

It is almost inconceivable to believe that all that is and all that was started from an immense release of energy in a space smaller than an atomic nucleus—smaller than even a proton. Having a clearer understanding of our origin and the origin of the universe does not make us any less spiritual, it only tells us that we must look deeper—beyond the creation of space and time—deep within ourselves.

The Big Bang model of the universe describes its creation from a singular point about ten billion years ago. From this cataclysmic event, all that exists came into being. Before the Big Bang there was no space. There was no time. Nothing existed of this universe. It is incorrect to conceptualize this nothingness as an immense, dark void. A void implies space, and existence in this void, as a passive observer, implies the passage of time. Up until then, there was no space to contain such a void, and there was no time. Space and time are the embodiment of the fabric of the universe. Before the creation of the universe there was nothing.

It is inappropriate to talk of space before the Big Bang, for it is from this cataclysmic event that space was created. For three-dimensional creatures, such as us, thinking of "space" and a moment before space existed causes confusion. Saying that space and time (hence spacetime) unfolded at the moment of the Big Bang describes the genesis event more accurately. Actually, three dimensions of space and one of time unfurled. The remaining spatial dimensions remained infinitely curled in on themselves. Current leading theories describe six additional dimensions that stayed infinitely small. This brings the total number of dimensions of the universe to ten.

What is a dimension? We normally take the concept of dimension for granted. We are three-dimensional creatures. We have length, width and height. A chair has three dimensions—it is so many inches long, so many wide and only so high. A dog has three dimensions: length, width and height, and so on. A two-dimensional "creature" is a square, a rectangle, a triangle, a circle. Two-dimensional creatures have length and breadth, but no concept of up or down. Their universe is flat. In a two-dimensional universe the circle sees the square from the same perspective—from edge-on. The circle sees the square as a line (and vice versa). Pretty dull, but they cannot perceive each other from above. Now visualize a one-dimensional creature—a line. The only distinguishing feature characteristic from one line creature to the next is length, for these lines have absolutely no width. Creatures of zero dimensions exist as only a point—no length, no width, and no height.

It is difficult, maybe even impossible, to visualize a fourth dimension. As three-dimensional creatures we know length, width and height, but what dimension lies beyond that lies beyond the grasp of three-dimensional humans (so far). The extension of a point (zero dimensions) is a line (one dimension). If a line is extended perpendicular to itself, a square (two dimensions) is formed. Similarly, if a square is extended perpendicular from itself, a cube (three dimensions) is formed. If it was possible to extend a cube perpendicular to itself simultaneously in all three dimensions, a four-dimensional hypercube would be formed. This is impossible to form or even conceptualize in three-dimensional space, so do not even try. However, a fourth coordinate or dimension is required to define the location of a specific event in space (or on earth). One must specify where—latitude, longitude, altitude (or length, width, height from a datum)—and when the event took place or will occur. Time is a fourth dimension. Future and past are directions, just like up-down and left-right. Time is a dimension, but uniquely different from the other three in that it has a direction. We move only into the future, never into the past. We can step left and right, forward and back, and even up and down, but we can move only in the future direction of the fourth dimension. We cannot (as far as we know) move back in time, into the past.

Space is a continuum—a spacetime continuum. Einstein viewed the universe as a continuum warped by gravity. He described gravity as a distortion of space and time. High intensity gravitational fields such as those emanating from stars and black holes distort, or warp, spacetime more severely than the gravitational fields produced by moons and planets such as the earth.

How do we conceptualize this dimension of time? Time is a manmade concept. In our mature universe of today, looking out into space is the same as looking back in time. Since light has a finite (although extreme) speed—approximately 300,000,000 meters per second (186,000 miles per second)—starlight that reaches our telescopes today came from an event in the universe's past.

Our closest stellar neighbor is in the Alpha Centauri system. This is a binary pair both with similar mass, diameter and luminosity of the sun. The binary pair is orbited by a tiny red dwarf, Proxima Centauri, in a harmonious ternary waltz. A weary traveler in a spaceship capable of speeds of a hundred thousand miles per hour (the space shuttle orbits the earth at approximately 17,000 mph) would arrive in this system after some 30,000 years have passed. The fastest machine ever sent out by mankind, the *Voyager 2* spacecraft, is traveling at approximately 50,000 miles per hour. Alpha Centauri is just over four light-years distant to us (4.3 to be more precise). A light-year is the distance in space that light travels in the time span of one year. This distance spans approximately six trillion miles (6,000,000,000,000). Light traveling from the Alpha Centauri system took a little over four years to reach our telescopes and the retina of our eyes. So, the event we are witnessing (starlight from Alpha Centauri) took place four years in the past.

As you can see, looking out into space is equivalent to looking back in time. The closest galaxy to our Milky Way, the Andromeda Galaxy—M31—is approximately 2.2 million light-years distant. Seeing this tiny, fuzzy spiral through binoculars on a dark night, you view two million years into the past. Light left this galaxy when humans first existed on this planet. Even our star, the sun, is eight light minutes away (approximately 93 million miles). The sunlight that reaches the earth has been traveling for eight minutes. If the sun were to implode at this very moment, we earthlings would be completely unaware for a few minutes because the light we see transpired just under ten minutes ago. Don't throw out your sunscreen just yet! Our sun has about five billion years of life energy remaining.

The universe is a vast expanse of space and time. Starlight that left a star in a distant galaxy hundreds of millions of years ago streams endlessly through time. The star may be long gone, an extinguished, dark cinder, but its light that left long ago keeps its memory alive for any consciousness that happens to intercept it.

EXPANSION

There was a time in the distant past—billions of years ago—when the universe was a much smaller place than it is today. Current estimates of the age of the universe put it somewhere between 10 to 15 billion years (more precisely, 14 billion years). Space is expanding. The fabric of the universe, the spacetime continuum, is spreading out in every direction simultaneously. The universe expands in all four dimensions (three of space, one of time).

Everywhere in the sky, past a few hundred million light-years, distant galaxies are receding away from our own Milky Way, and these galaxies are moving away from each other as well. As the universe expands, every distant point in space moves away from every other point. Just as raisins in a loaf of raisin bread all move away from one another while baking, giant clusters of galaxies anywhere in the sky are becoming more and more distant.

The distant recession of galaxies, and hence the expansion of the universe, is evident in the redshift of galaxies. The path of the light waves being stretched as the light travels on its long journey causes the redshift of distant light sources (galaxies). The spectrum of light is shifted down in frequency toward the red portion of the electromagnetic spectrum. Light is simply electromagnetic energy waves. Since light travels in waves, we can associate the energy of the wave with its frequency. Visible light in the red portion of the electromagnetic spectrum is lower in frequency than blue light. If a galaxy was moving toward us—very few are—its light would be shifted toward the blue portion of the spectrum. This phenomenon is due to the Doppler effect.

The Doppler effect is also evident in sound waves. There is a sudden change in pitch (from high to low) when a speeding ambulance passes. Sound waves being compressed in front of the ambulance produce a higher pitch. Similarly, sound waves expand behind the ambulance causing a lower pitch to observers left behind. Galaxies in the Local Group do not appear redshifted to observers on earth due to local gravitational tides holding the group together.

Edwin Hubble, in 1929, studied the distant redshift of galaxies. Hubble, for whom the space telescope is named, attributed this shift in frequency of the light from distant galaxies to the expansion of the universe.

Galaxies were previously thought of as spiral nebulae within our own island of stars. Hubble proved that entire galaxies of stars exist far beyond our own Milky Way and, in fact, these so-called nebulae were distant and separate from our own galaxy. He went on to describe galaxies as the basic unit of distribution in the universe. Hubble noticed how galaxies are distributed in distance—that they increase in number proportional to the volume of space surveyed. The farther an observer looks into space and back in time, the more galaxies he or she will see. The universe before Hubble was a tiny, humble place in mankind's eye. Einstein laid the foundation for the intricacies of space, time and gravity. Hubble extended the knowledge of how vast the universe is and verified Einstein's prediction that it is getting bigger. The universe is expanding. Sometime in the distant past the universe must have been tremendously more dense and hot.

The Hubble Law

Hubble was instrumental in correlating the redshift in the spectra of galaxies with their distance. This is the Hubble law:

> The farther away a galaxy is, the greater its redshift will be.

The Hubble law still holds true today, verified with thousands of galaxies using modern day telescopes and equipment.

Hubble located Cepheid variable stars in distant galaxies. Cepheid variables are stars that brighten and dim over intervals of time, and whose cycle is directly related to their intrinsic brightness. The brightness of a star is indirectly proportional to the square of its distance (from Newton). Therefore, the distance to a star, such as a Cepheid variable, whose brightness is known, is easily calculated. Cepheid variables are used as the "survey markers" of the heavens.

Hubble was able to correlate the Cepheid variable star's distance with the galaxy's redshift. The distance calculations agreed with amazing accuracy. Hubble found that he could estimate the distance of galaxies by measuring the brightness of the Cepheid variables contained within them. He found a linear relationship between the distance of a galaxy—measured by the brightness of Cepheid variables—and the galaxy's redshift. Hubble concluded that all distant galaxies are receding from us and from each other as well. By correlating the amount of redshift with distance (the more distant the galaxy, the more evident its redshift), Hubble found the cosmic expansion that Einstein predicted, but was hesitant to admit. Hubble provided undeniable proof that the universe is expanding, and the galaxies (the basic unit of distribution in the universe) are riding within the very fabric of spacetime that is expanding.

The expansion of the universe is evident in the redshift of distant galaxies. Light from distant galaxies is stretched out along with the fabric of spacetime. Their ancient light is shifted toward the red end of the spectrum, into longer wavelengths. The Hubble law explains that the more distant a galaxy is, the faster away it will be receding. The Hubble law may be stated in equation form:

$$v = H_0 d$$

where v is the velocity of a distant galaxy, d is its distance and H_0 is known as the Hubble Constant. Evident from this equation, a galaxy's speed of recession (velocity) is directly proportional to its distance.

The Hubble Constant is a description of velocity per unit distance. Current estimates place H_0 somewhere between 50 and 100 $km \cdot s^{-1} \cdot Mpc^{-1}$ (kilometers per second per megaparsec). A parsec (megaparsec = one million parsecs) is an astronomical unit of distance. A parsec (or parallel second) is the distance to a star that forms a triangle between the sun and the earth equal to one second of arc.

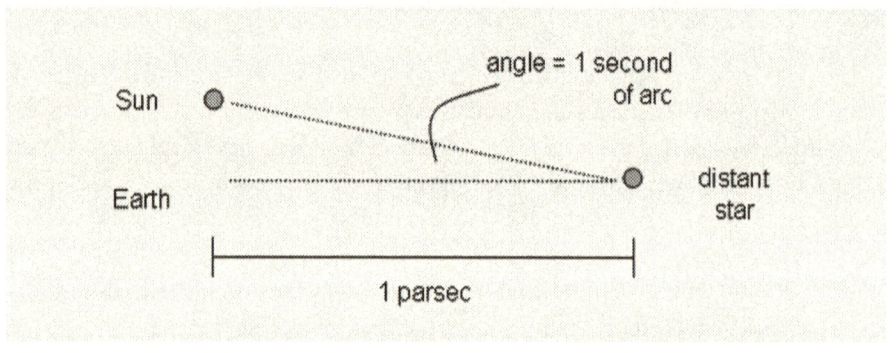

A parsec is a unit of distance equal to about 3.26 light-years. It is the distance to a star that makes 1 second of arc with the sun and the earth.

One parsec is equivalent to 3.26 light-years, so a megaparsec totals a distance of 3.26 million light-years. Therefore, a Hubble Constant equal to 100 $km \cdot s^{-1} \cdot Mpc^{-1}$ equates to a universal expansion rate of 100 kilometers per second (velocity) per 3.26 million light-years (distance). For, example, a galaxy that is 3.26 million light-years away (one megaparsec) should be traveling away from us at a velocity of 100 km/s.

Recent estimates of the age of the universe have produced conflicting results. Estimates of the age of certain stars with known mass, luminosity and chemical composition have conflicted with the predicted value of the Hubble Constant making these stars older than the estimated age of the universe. Obviously, this discrepancy presents a paradox. No star can be older than the universe itself. How can a child be older than his mother?

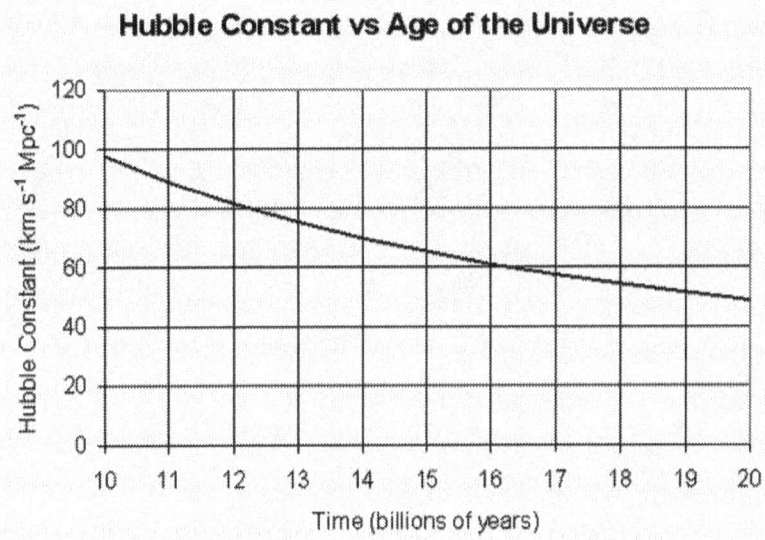

Hubble's law states that the velocity of a distant galaxy is proportional to its distance. The Hubble Constant (H_0) is a description of the expansion rate of the universe per unit distance. The distance light travels is a measure of time since the speed of light is constant. An accurate measure of H_0 should give an accurate prediction of the age of the universe.

The difficulty lies with the Hubble Constant itself—it is not really a constant. It is a description of the current expansion rate of the universe. It is constant only if the expansion rate of the universe has always been unchanging. This notion is highly unlikely due to the consistent pull of gravity between galactic clusters. The eternal pull of gravity tends to slow expansion. A recently conceived hypothesis hints of an undetected repulsive force speeding up the expansion of the universe, but this idea is highly speculative.

The point is that the Hubble Constant is difficult to estimate. The goal is to find a value consistent with the age of old stars in nearby global clusters and the age of distant galaxies. At that time we will have a more accurate value for the age of the universe itself.

Today's universe is unquestionably expanding. Galaxies and clusters of galaxies are moving farther and farther apart. This fact follows the notion that sometime in the past, clusters of galaxies were closer together than they are at present, and even closer still in the distant past. This notion of expanding space leads to the idea that all matter in the universe (matter makes up galaxies, people and puppies alike) originated from a singular explosive point in spacetime—the Big Bang. What lies beyond that event is unknown. Obviously the event (the Big Bang) occurred, but why? What mechanism initiated such an event? As Steven Hawking put it, there is no reason for a beginning. Some force or entity must have initiated such an event. Hawking feels that to understand this is to know the mind of God.

CURVATURE OF THE CONTINUUM

The general theory of relativity describes the curvature of spacetime. The theory explains how this warping of space is a direct influence of mass. All matter contains mass, and all mass is influenced by gravity. The more massive an object is, the stronger its gravitational attraction to other bodies. General relativity models the universe warped by massive bodies, and it is this indentation or depression in the fabric of the continuum that grips other objects and attempts to pull them in.

Einstein was convinced that the universe was static, neither expanding nor contracting, but his general theory of relativity did not support the static universe hypothesis. To account for a flat, unchanging universe he modified the theory and introduced into his equations of gravity what he called a "cosmological constant." This term counteracted the natural tendency for a homogeneous (uniformly distributed) and isotropic (same in all directions) universe from collapsing under its own gravity. Einstein's cosmological constant acted as some kind of repulsive force opposing gravity and preventing the universe from collapsing.

Einstein later refuted his static universe theory along with the cosmological constant term stating that it was "the greatest blunder of his life." Einstein's general relativity correctly predicted a dynamic, changing universe, but he modified this prediction to produce a flawed, static model of the universe. In 1929 Hubble found unmistakable proof of an expanding universe in the consistent and unanimous redshift of distant galaxies.

Assuming the cosmological principle of a homogeneous and isotropic universe, three possible geometries exist for the universe as a whole: open, closed and flat. The simple geometry lessons learned in school are based on Euclidean geometry. This geometry is flat. The shortest distance between two points in Euclidean space is a straight line. The sum of all three angles of any triangle is 180°, and the sum of all four angles of a square sum to 360°.

When discussing the geometry of the universe a key parameter must be considered: the density parameter of the universe, Ω (omega). The density of any object is simply how much "stuff" is contained in the object or, more precisely, the amount of mass per unit volume. The density parameter of the universe, Ω, is the ratio of the average density of the universe to the critical density. The critical density is that density that is just enough to eventually halt the expansion of the universe but insufficient to cause its collapse. Again, there are three possible geometries for the universe: closed, open and flat.

First, consider a closed universe. A universe that embodies a closed geometry will continue to expand, eventually halt and begin to contract. Within this universe the mass density is high enough ($\Omega > 1$) to halt the expansion of the continuum via the force of gravity. Gravity will maintain its hold and begin to contract the universe, gripping and squeezing tighter and tighter.

The Big Bang, by this time a distant memory from many eons ago, is now a Big Crunch. The universe shrinks smaller and smaller. Clusters and superclusters of galaxies (now blueshifted) are moving closer and closer together. Perhaps the destiny that awaits a closed universe remains with an infinitesimal point from which space and time unfurled—a cosmic full circle. Would this lead to another Big Bang? Perhaps existence is nothing but an eternal series of oscillations of creation and destruction, each cycle lasting tens or hundreds of millions of millennia.

The density of a closed universe causes spacetime to curve back onto itself. As three-dimensional creatures we cannot fathom its appearance, but the best analogy is the universe depicted as the surface of a sphere. The center of the sphere depicts higher dimension. Here, the shortest distance between two points on the surface of the sphere is not a straight line, but an arc. A triangle contains more than 180° and a square more than 360°. If the two-dimensional inhabitants of the surface of the sphere could somehow find a shortcut through the center to get to a distant point on the other side, they would find a shortcut through their universe. Here lies the concept of interesting space travel: a shortcut is found by warping through a higher dimension. For now, this is science fiction, but one can only hope!

Next, consider an open universe. A universe that embodies an open geometry will continue to expand forever. Within an open universe the mass density is too low ($\Omega < 1$) for gravity to cease expansion, so the universe continues to expand farther and farther without bound, creating a cold, dark and bleak existence for the distant future. In this universe, galaxies will continue to recede, and their stellar constituents will eventually burn out into dead, smoldering cinders—a lot of which will be consumed by their central gigantic black hole. The universe will become a dark and dismal place. Galaxies will be mostly dark swirls of ash with the faint glow of white dwarfs. The dark and lonely future of an open universe makes the certain death of the Big Crunch of a closed universe a welcomed friend.

The density of an open universe causes spacetime to curve away from itself. The geometry is analogous to a saddle or hyperbolic space. The shortest distance between two points is not a straight line, but an arc. The angles of a triangle in an open universe sum to less than 180°, and a square similarly contains less than 360°.

Now consider an expanding universe whose mass density is just enough to eventually stop the expansion, but there is not enough mass to collapse the universe. The dynamic universe eventually becomes static. The geometry of such a universe is flat. The mass density of a flat universe is equal to the critical density ($\Omega = 1$), so expansion is stalled, but collapse is averted. The geometry of a flat universe is Euclidean, where the shortest distance between two points is a straight line, triangles make 180° and squares make 360°.

This flat geometry is unstable though. Its mass density accomplishes a knife-edge balance between inevitable collapse and eternal expansion. An unlikely scenario, but this seems to be the case for our universe. This is known as the *flatness problem* (see the section on Inflation).

The amount of mass in the universe will determine its fate. Mass exists in many forms from the exotic to the mundane. Visible stars account for only a tiny fraction of the mass required to close the universe. Matter more difficult to see such as brown dwarfs, red dwarfs, white dwarfs and black holes may be accounted for by their affect on surrounding material and the fabric of spacetime itself. Perhaps unseen and barely detectable particles—neutrinos (little neutral one)—that permeate the universe and outnumber the matter particles by a staggering amount, contain the least bit of mass where they were once thought to be without mass. Perhaps our equations of gravity and spacetime are not as precise as we think, and we have yet to conceive of some anomaly.

FUNDAMENTAL FORCES

There are only four fundamental forces throughout the universe.

During the Planck epoch when the universe—all that is and all that used to be—was much less than a second old, energy existed in a very exotic form. In fact, all that existed was energy. The entire universe existed as an infinite energy well in an extremely minute space. To understand the extreme conditions of the embryonic universe, consider how energy is related to force. Force, by definition, is the net result of an accelerating mass (Newton's Second Law): $\Sigma F = ma$

Energy, or, more appropriately "work" (from basic physics), is a force realized over a distance.

All of the forces in the infant universe were united into a single "superforce." This unification of forces defines an era when the fundamental forces were described by a single equation. If the relativistic equation, $E = mc^2$ unleashed the power of the atomic nucleus, fathom the meaning of a "theory of everything," where the fundamental forces of the universe are described by a single equation. It may seem only academic, but such knowledge would indeed be the greatest scientific discovery of all time. Perhaps truly unlimited sources of clean energy would result, forever feeding the world's energy requirements. Perhaps engines could be built to take us to the stars in only an instant. On a darker note, perhaps our ultimate discovery would be our demise. Knowledge is power. As with all scientific discovery we must continue to explore and expand our knowledge, but it is imperative that we proceed with caution. The lessons learned from our carelessness now will become exponentially costly.

Nature has constructed herself upon four fundamental forces. That is it, only four. No matter where your mind may take you from the chair you are sitting in to the event horizon of the multimillion solar mass black hole that exists at the center of our galaxy, there are only four forces that interact with matter. Each force is carried by a particular quantum (particle). The four fundamental forces of the universe are as follows (from weakest to strongest):

1. Gravity (carried by gravitons)

2. The weak nuclear force (carried by W and Z bosons)

3. The electromagnetic force (carried by photons)

4. The strong nuclear force (carried by gluons)

The strong and weak forces are effective only at extremely short range. Gravitational and electromagnetic forces extend their reach to limitless distances across the galaxy and even across the universe. The strong, weak and electromagnetic forces are responsible for atomic structure—the building blocks of matter. Gravity is responsible for the large-scale structure of the universe: planets, stars, solar systems, galaxies and supercluster complexes of galaxies.

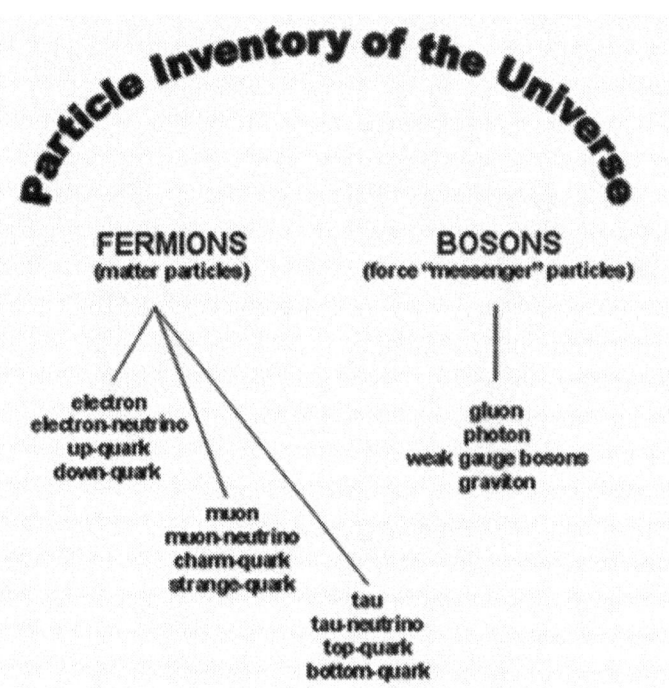

According to the Standard Model the universe is composed of matter particles (fermions–quarks and leptons) and messenger particles (bosons). Further division of the particulate universe may lead to perhaps the ultimate building blocks—strings.

Electromagnetic Force

The careless observer may inquire about the force his hand produces pushing down on the tabletop, perhaps even labeling this as the "brute force." "Since matter is composed mostly of empty space," he asks, "why doesn't my hand go right through the table?" Well, the table pushes up on our observer's hand with an equal (and opposite) force to that which he uses to push down on the table. This is Newton's Third Law:

> For every action (force) there is an equal and opposite reaction.

The force our foolish observer feels from the table pushing back is attributed to the electromagnetic force. Our hands—flesh, bone, muscle and tendons—are composed of atoms and molecules just like everything else. The positively charged atomic nucleus keeps its negatively charged electrons in orbit, just as our sun keeps her planets in continuous orbit. The opposite charges of these two types of particles (electrons and protons) are what hold the atom together. Opposite charges attract, while like charges repel. The electrons in the outer shells of our observer's palm are repelling the electrons in the outer shells of the tabletop's surface. The harder he pushes down, the greater the repulsive force between the electron shells. So, there is no "brute force," only the electromagnetic repulsive force between two objects.

Electromagnetic radiation exists all around us. Radio waves, microwaves, infrared, visible light, ultraviolet, x-rays and gamma rays are all integral components of the electromagnetic spectrum. The electromagnetic force is carried by quanta named photons, and becomes weaker with distance but has limitless range.

Photons exhibit both particle and wavelike phenomena. Light, for example, (as well as radio waves microwaves, x-rays and the like) may be thought of as particles with wavelike properties—or waves that behave like particles. The wavelike properties of light and radio become obvious when dealing with the frequency of radio waves or the wavelength of light (color).

Gravity

Gravity is the force for which we are the most familiar. It holds us down in our seat, keeps pens, pencils, books and the dog from floating around the room. Gravity keeps the moon in orbit and the planets in orbit around the sun. Gravity holds galaxies and groups of galaxies together. It is the force that caused a gigantic cloud of gas, containing mostly hydrogen and traces of heavier elements—carbon, oxygen, nitrogen, iron—to condense some five billion years ago to form our solar system.

Every particle of matter—every atom—attracts every other particle of matter. Two atoms in the vacuum of space, even separated by millions of miles, are attracted to each other by the force of gravity. This attraction is extremely weak, but it is present.

Today, science tells us that gravity has infinite range. Newton taught us that this force is proportional to the square of the distance between two massive bodies. In other words, the moon, gravitationally bound to the earth, is about 250,000 miles distant. If this distance was halved (125,000 miles) the force of gravitational attraction between the earth and moon would be four times greater. High tide would have a whole new meaning! Consequently, the moon would be forced to orbit the earth twice as fast to counteract the gravitational attraction and prevent the two bodies from crashing into each other. Earthlings would witness a full moon every two weeks! There would be twenty-six new moons a year, but the ensuing climatic changes would make our home a violent place to live.

Gravitational tides would stretch and pull at the surface causing massive volcanic eruptions and earthquakes. The high tides occurring four times a day would stretch the earth almost into an egg. The entire eastern seaboard, as well as every other coastal region of the world, would be uninhabitable, for the high tides would come inland for perhaps several hundred miles (depending on the flatness and elevation of the land).

That is enough of global catastrophe for now. The point is that gravity, although the weakest of nature's forces, plays a vital role for our very existence. Gravity becomes weaker with distance but, like the electromagnetic force, has limitless range. Every particle of matter (atom) is attracted to every other particle of matter regardless of size and distance. Because of its weakness, the affect of gravity becomes evident only when dealing with masses the size of moons, planets and stars.

Strong Force

The strongest of the fundamental forces—the nuclear strong force—holds the nucleus of the atom together. Having like charges, protons naturally repel each other in response to the electromagnetic force. Luckily, for the sake of the universe, the strong force (boring, but aptly named) is a hundred times stronger than the electromagnetic force. The electromagnetic force, as well as gravity, has limitless range, but the strong force acts at only a very short range (approximately 3×10^{-15} m). The energy required to make or break a bond sewn together via the strong force exists inside the core of stars, where nuclear fusion rages and immense stores of energy are released.

The incredible amount of energy released from a nuclear weapon, or a more benign nuclear reactor, results from the emancipation of the nuclear strong force in nuclear fission reactions. Only a small amount of fuel is required to release a huge quantity of energy. This reaction, unlike burning coal and natural gas, is very efficient. A portion of the fuel matter (uranium and plutonium) is converted directly into energy. Remember $E = mc^2$? For this reason the Trident-class nuclear submarine, requiring literally only a handful of fuel, may stay out to sea for months without refueling. The only reason that brings the submarine back to port is the stamina of the crew—family and food supplies.

As previously stated, the strong force is responsible for containing the protons and neutrons of the atomic nucleus. The force also contains the particles (protons and neutrons) themselves. Modern physics has taught us that protons and neutrons are not fundamental. They are not elementary. Protons and neutrons are indeed made of smaller quanta—quarks. Two "up" quarks and one "down" quark compose a proton. Conversely, two "down" quarks and one "up" quark constitute a neutron. The "up" and "down" quarks have different quantities and opposite polarities of electric charge. Electrons have a charge of –1. Protons exhibit a charge of +1, while neutrons have no charge, 0. An "up" quark has a fractional charge of +2/3, and a "down" quark has a fractional charge of –1/3. These charges summed together result in a whole number charge for protons (2/3 + 2/3 + -1/3 = +1) and neutrons (-1/3 + -1/3 + 2/3 = 0).

These three particles are also held together by particles called gluons—the carriers of the strong force. Today, even quarks are not thought to be elementary. They are thought to be composed of tiny vibrating strings. These strings could perhaps be the remnants of the fetal universe—the higher remaining dimensions that failed to unfurl.

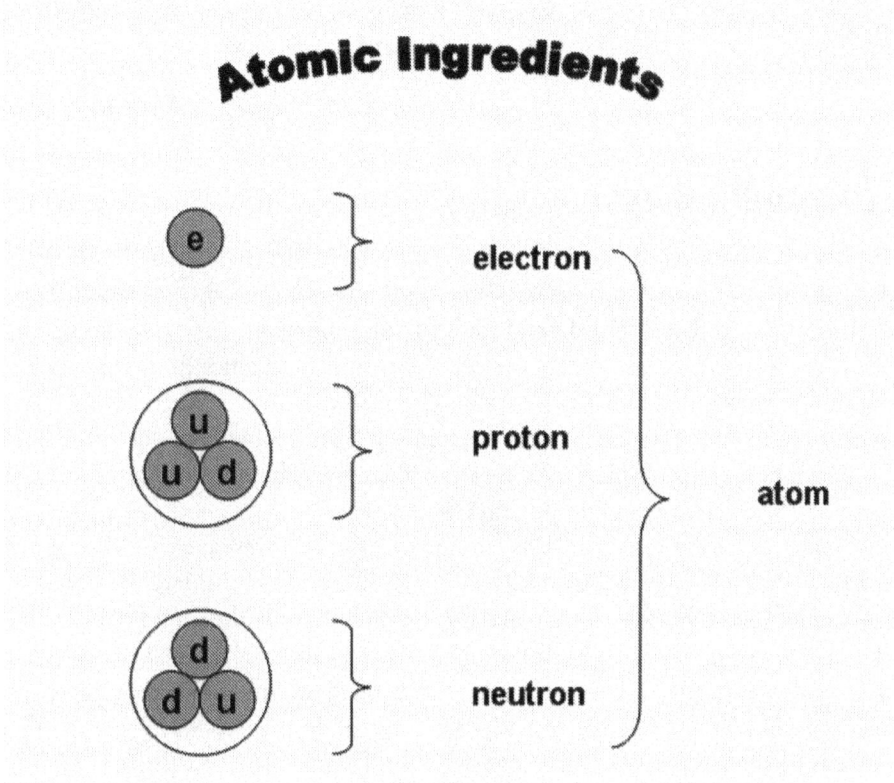

Atoms are composed of a nucleus (containing protons and neutrons) orbited by electrons. Protons, neutrons and electrons are members of a family of particles called baryons. Protons are further divided into 2 up-quarks and 1 down-quark. Neutrons are composed of 2 down-quarks and 1 up-quark.

Weak Force

The nuclear weak force is responsible for radioactive decay and the production of neutrinos. The weak force is carried by particles called "weak gauge bosons"—the W and Z bosons. The nuclear weak force is transmitted when one "down" quark within a neutron spontaneously transforms into an "up" quark. This action produces a proton where a neutron once existed and transmits a W-boson. The W-boson almost immediately decays into an electron/antineutrino combination or a positron/neutrino pair. The transformation of a neutron into a proton by the transmission of a W-boson is referred to as beta decay.

Nature does not favor free neutrons. A stand-alone neutron will decay into an electron, a proton and an antineutrino in a matter of minutes:

$$n \quad \rightarrow \quad e^- + p + \bar{\nu}_e$$

Quantum Filaments

Two of the leading theories of cosmology, the general theory of relativity (encompassing the very large) and quantum mechanics (engrossing the very small), are, to date, the best theoretical description of nature; however, they are incomplete. They cannot both be absolutely correct. Quantum mechanics does not account for gravity, and general relativity cannot account for miniscule fluctuations called "quantum foam," predicted within extremely microscopic areas of spacetime. Both theories are incomplete and need modification. The "superstring theory" of the mid-1980s attempts to do just this.

Atoms (from the Greek term meaning "indivisible") were recently thought of as being the smallest constituent of matter. The atom was considered the smallest particle of any substance. Soon the fog was lifted from the cloudy shell. The clouded electron shell around the atom was found to be composed of tiny electron particles. This discovery heralded the birth of modern chemistry. The encompassing cloud was penetrated, and the atomic nucleus was exposed. The nucleus of the atom soon revealed its secret and exposed its constituent building blocks of matter: protons and neutrons. For awhile, these conceivably impenetrable, point-like particles were the material building blocks of matter: electrons, protons and neutrons. Soon, however, the nucleus was smashed in high-energy particle accelerators obliterating protons and neutrons into their constituent particles, quarks. Each proton and neutron was found to exist as a ternary waltz of three quarks each. Some modern physicists and cosmologists theorize now that quarks and electrons, as well as the force-carrying particles, such as the photon, are not elementary but that they are each composed of a tiny filament of space-time itself, referred to simply as a "string."

According to superstring theory, the smallest constituents of matter are not point-like particles—like quarks, leptons and photons—but tiny, vibrating filaments of strings. The strings are one-dimensional loops like rubber bands on the order of the Planck length (10^{-35}m), much too small to probe with even modern technology. The theory suggests that each familiar particle (quark, lepton, photon, gluon, etc.) constitutes one string. The bold and extremely complicated theory proposes that all matter and forces are composed of oscillating strings.

According to the theory, strings can undergo any number of resonant frequencies or vibrational patterns like a plucked rubber band stretched between fingers.

A guitar string that is plucked will vibrate at a certain frequency depending on the tension in the string. The vibrational pattern of the guitar string gives rise to a whole number of wave crests produced on the string, and emanates a tone or specific musical note. That musical note is characteristic of the string relative to its tension. The "musical notes" played by the strings of superstring theory that proliferate all matter and energy, reveal themselves as the different masses and charges of the different particles. Different modes of string vibration describe different masses and force charges. A particle's mass and charge are dependent on the resonant frequency of the string in which it is composed. Particles with greater mass have vibrational patterns of higher frequency and higher amplitude and hence higher energy.

Each and every elementary particle (quark, electron, photon, graviton, etc.) is composed of a single string. All strings are physically identical. Only the resonant vibration will vary between differing particles. Of what, then, are the strings themselves composed? Perhaps a more precise description would be "quantum filaments" or "cosmic threads." Just as particles of matter are trapped packets of energy (since matter and energy are different manifestations of the same entity), perhaps these strings are trapped dimensions of spacetime that did not unfurl at the Big Bang. Perhaps these strings, the building blocks of matter and energy, are in fact curled-up higher dimensions of spacetime.

Space and time were created at the Big Bang—three spatial dimensions and one temporal dimension unfurled from an infinitesimal point. It is conceivable that extra higher dimensions (inconceivable to three-dimensional creatures such as us) remained curled in on themselves and failed to expand with the fabric of spacetime. Think of an unfurled dimension as a sheet that is rolled into a tight roll—into a tube—and then spiraled around lengthwise into a tight roll. If this sheet were to unroll in only the lengthwise direction, like the expanding universe, it would still have an extra dimension tightly curled in on itself. The actual mechanics and mathematics of hidden, higher dimensions are much more complicated than this simple example, but the concept remains the same.

Superstring theory predicts that the universe has as many as six higher dimensions woven into the fabric of the universe that did not unfold at the Big Bang. Perhaps these strings—the building blocks of matter particles and messenger particles—are the manifestations of these compact dimensions. It would be ironic, yet magnificently beautiful, if all matter and energy in the universe were indeed constructed from the very fabric of the universe itself!

THE FIRST FEW MINUTES

In just a fraction of a second space, time, matter and energy were born. This momentous event transpired some 10 to 15 billion years ago as the universe was created. In this instant in time, ten dimensions came into being—four of which unfurled: three of space and one of time. All that is and all that was resulted from the explosive expansion and rapid cooling of the primordial soup of radiation and matter. The colossal structures evident today, from galactic clusters to supercluster complexes of galaxies, as well as all matter on the earth—from Mt. Everest to the creatures dwelling in the deepest, darkest sea—came from that 100 billion-degree primordial soup born in that instant in time.

It is uncommon when studying and describing the universe to use timescales of a billionth of a trillionth of a trillionth of a second (10^{-34} seconds). We are accustomed to measuring in the millions and billions of years. The universe of today is a static place (only seemingly so) where meaningful events on the macro scale transpire over hundreds of millions of years. However, when peeking into the obscurity of the first few moments of creation, when the universe explosively expanded to a sizable portion of its current greatness, we are dealing with timescales comparable to a phone conversation or a coffee break. Ponder this—if you had gotten up to take a break during the coming attractions of this grand feature presentation, you would have missed the entire creation. Perhaps you would have heard a tremendous explosion.

During the first few microseconds of creation, the temperature and kinetic energy of the surrounding energy flux is far too intense for atoms and even nuclei to be bound together. Not even protons and neutrons can exist. The universe is swimming in a sea of quarks. Most of the energy density of the primordial universe is contained in background radiation. This background radiation, in the form of energetic photons, can still be detected today, some ten billion years later, as the cosmic background radiation. The violent temperature of this ancient energy soup has cooled somewhat. The cosmic thermometer now reads about 2.7 degrees above absolute zero. This radiation, now shifted down to the microwave region of the electromagnetic spectrum, still echoes throughout interstellar space.

It is a feeble reminder woven into the fabric of space and time of a once violent past.

The cosmic background radiation was discovered by accident, in 1964, by two radio astronomers at the Bell Laboratories in New Jersey—Arno Penzias and Robert Wilson—who, in 1978, won the Nobel Prize in physics for its discovery. They were developing a new type of horn antenna to detect radio signals from orbiting satellites—specifically the recently launched *Echo I* satellite.

From any direction in the sky to which they pointed the antenna, they detected a constant noise. They attributed this annoyance to receiver noise and messy pigeons inside the antenna horn, but the noise persisted. Many great discoveries are encountered by accident, and this is certainly no exception. The existence of background radiation emanating throughout the cosmos and its amazing homogeneity (a little more than one part in a million) is a major supporting pillar in the Big Bang theory. This discovery brilliantly confirmed the theory of a universe that began existence from a Big Bang and produced an afterglow of radiation that persists some ten billion years later. The overall background radiation does not vary by more than a few thousandths of a degree no matter which direction into deep space our telescopes peer. This was proven in the early 1990s by the COBE satellite (Cosmic Background Explorer).

Back to the first few moments of creation…

In the intense fireball of our young home, matter existed in the form of free quarks and antiquarks (the antimatter equivalent of a quark). Free quarks are a rarity of nature, at least in today's universe. Quarks come in triplicate. The strong force is not strong enough to maintain the bond between three quarks (or antiquarks) in the intense heat of this "quark soup," so solitary quarks and antiquarks are free to do their dance. Matter and antimatter do not mix. They are bitter enemies, with no hope for peace. When a matter particle meets an antimatter particle, or vice versa, they destroy each other in total and complete annihilation. Their total mass is converted to energy in the form of gamma rays—photons of the most extreme energy level. However, for reasons unknown to present-day science, there is a very slight excess of matter over antimatter, so a tiny fraction of matter (quarks) survive. This tiny fraction constitutes all matter that exists today in the universe.

The remaining quarks begin to condense into protons (whose current tally is estimated at 10^{78} in the universe) and neutrons. This number is huge: approximately 1,000,000,000,000,000,000,000,000,000,000,000,000,000,000,000,000,000,000,000,000,000,000,000,000,000,000,000,000,000 protons in the visible horizon of the universe! Protons are not created in stars. All matter (protons) that exists in today's universe came from this primordial excess of matter over antimatter when the universe was only a few microseconds old. After a few microseconds, no solitary quarks remain. All quarks in the primordial soup found two other lonely quarks and formed a union in triplicate. The union of quarks formed the protons and neutrons.

The universe is now one second old.

The universe is totally dominated by photons. The photons in this radiation bath are extremely energetic. In the unbelievable heat and pressure of the plasma, some protons and neutrons fuse to form deuterium (heavy hydrogen) and then helium. A great deal of the entire matter inventory (approximately 23 percent) of the universe is transformed into helium and a tiny amount of lithium.

Welcome to the era of nucleosynthesis. The fraction of helium synthesized (23 percent) remains evenly distributed in today's universe, and astronomers can account for this percentage of hydrogen and helium now in starlight. This consistent percentage of helium throughout the universe and its even distribution forms a basis for the Big Bang model of the universe. This process of fusion to form helium is halted after only a few minutes. The universe at this stage does not remain hot enough for heavier nuclei to form, due to its rapid expansion. The spatial expansion quickly cools the infant universe.

The universe is now a few minutes old.

The universe after the first few minutes experiences no more significant milestones for the next few hundred millennia or so. During this time the universe continues to be an intensely hot and extremely energetic place. The expansion of the universe progresses at a rate comparable to today's expansion rate. Particles of light (photons) cannot travel significant distances without colliding with particles of matter: electrons, protons (nuclei of hydrogen), deuterium and nuclei of helium. Photons during this era are quickly absorbed by the matter with which they are in constant collision. Immediately after absorbing a photon the matter quickly relinquishes the newly acquired energy and emits another photon. Due to the photon's inability to travel very far, the universe is opaque to light. Light cannot travel far through this ancient universe. If some sort of ancient astronaut were able to survive in the plasma and energy storm of this early universe, she would not be able to see very far. Perhaps it would be similar to being completely surrounded by a gigantic bubble bath—bright, white all around, but one would be completely unable to see anything. However, the immense density and pressure that persists would cause her extreme discomfort.

The universe is now a few hundred thousand years old. This is the era of photon decoupling. The temperature, at about 3000 degrees, is slightly less than the surface of the sun. The universe is expanding at the more leisurely rate observed today. The matter content of the universe is unchanging. The density of a substance is simply its mass divided by its volume. Since the volume of the universe is increasing, due to its expansion, its density decreases. The universe is becoming less dense. After about 300,000 years the universe has lost sufficient density so that photons may travel through the spacetime continuum relatively unimpeded. This plane of existence for the universe is called photon decoupling.

The universe now is relatively transparent. There is still not a great deal to see—just a seething plasma of tiny particles. The photons that are permitted to escape during the era of photon decoupling are "visible" today as the cosmic microwave background radiation. These particles of energy have lost significant energy and have been redshifted far down the spectrum, but they still exist today.

Seeing these photons today, we see the young universe when it was only a few hundred thousand years old. We may not "look" farther back than that. The stage prior to photon decoupling is impenetrable, impervious to photons. The cosmic background radiation observed today provides a snapshot of the universe at an age of about 300,000 years.

The violent fires of the infant universe are beginning to dwindle somewhat. Matter is now able to exist as highly energetic, charged particles—plasma. These are exciting times! After a million years since creation, the first atoms are formed. Protons (hydrogen nuclei), deuterium and helium nuclei are now able to capture free electrons to form atoms.

The universe is now a million years old.

Besides the energetic nature of the universe, it is still a rather boring place. Matter has not yet begun to conglomerate to form interesting structures like galaxies, planets and people. The universe will have to wait a hundred million years or so. The universe is basically a vast, dark expanse of hydrogen and helium gas for the next several hundred million years.

The universe is now a billion years old.

The Milky Way Galaxy has begun to form.

The universe is now ten billion years old.

Life on earth has gained the ability to ponder its own existence.

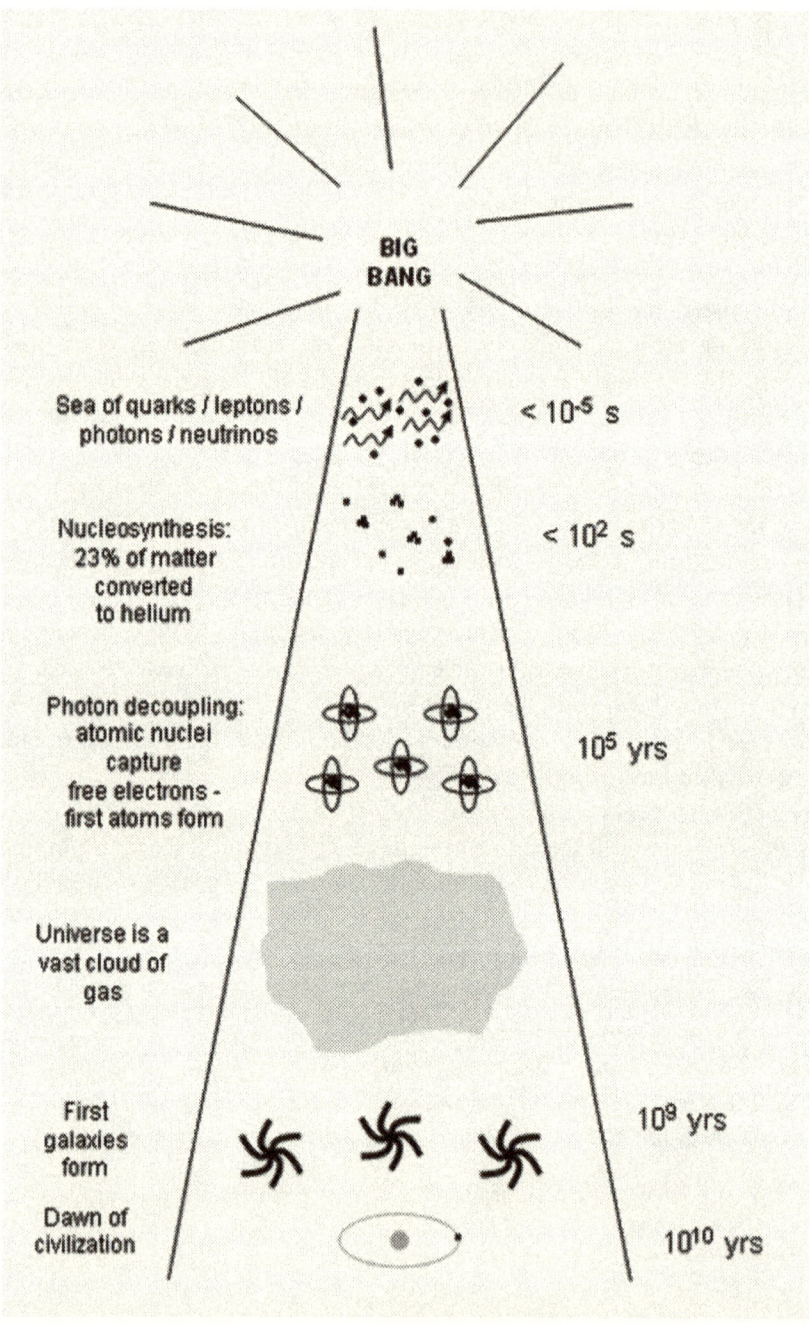

Timeline of the universe.

FORMATION OF THE SOLAR SYSTEM

If it were not for gravity, the universe would be a rather ordinary place consisting mostly of hydrogen and helium (as far as matter is concerned). There would be no galaxies, no solar systems. There would be no stars in the heavens, no planets, no earth, no pet dogs or cats, no us! If gravity failed to exist, the inflation of the universe would not have fallen prey to the initial minuscule density fluctuations (discovered by the COBE satellite in the early 1990s). It is from these density fluctuations so long ago that matter (mainly hydrogen, approximately 73 percent, and helium, approximately 23 percent) clumped together to form the seeds of these gargantuan supercluster complexes—giant walls of galaxies, evident in the large-scale universe of today. These vast strands and walls of galaxies are thought to be the largest structural formations in nature.

As galaxies formed from the seeds of ancient quasars seen billions of light-years distant, clouds of hydrogen and helium gas condensed and began to spin. Spin? (It is hard to keep anything static in this universe). The vast cloud of gas eventually became able to support itself against its inevitable collapse by the initiation of nuclear fusion. A star is born.

This process takes countless millions of years, of course. For a moment think back to the vast cloud of interstellar gas that existed before the solar system. The cloud consisted mainly of hydrogen and helium (approximately 99 percent). All elements contained in the present-day earth existed here in this vast cloud. Two atoms of hydrogen (one proton orbited by one electron) in the vacuum of space are attracted to each other by the force of gravity. Their masses are extremely tiny (but existent) and the force of attraction is immeasurably minute (but present indeed). As a result of this force of attraction, the two particles—in this vast cloud of interstellar gas—eventually begin to move toward each other. Sometimes, a lot of times, the two particles will collide and bounce off each other sending each on a new trajectory. However, many times the particles will miss and pass each other. At the moment of their mutual passing the distance between the two atoms will be the least and, hence, the force of attraction the greatest. Remember Newton's law of gravity?

> The force of gravity is inversely proportional to the square of the distance between two bodies.

This increased force between the particles at their minimum distance, in a sense, reaches out and causes them to move toward each other slightly as they pass. This motion is similar to the slingshot effect that spacecraft exhibit when brought close to a planet or moon on their way deeper into the solar system. As a result, the spacecraft accelerates and gains noticeable velocity. Gravity grabs hold and slings the mass onto a new parabolic trajectory. Similarly, this "near miss" sends our two friends on a new trajectory that eventually changes their straight-line paths into a sloppy orbit around each other (assuming no other particles intervene).

The sum of all motion (including the Brownian motion exhibited in fluids), for the most part cancels out in our gaseous cloud, but some direction of rotation wins out. Nearby supernovae (exploding stars) can send shock waves through the cloud, compressing and accelerating the gas. The shock wave, depending on its location relative to the cloud of gas, may also influence its rotation. The death of a nearby star breathes life into a lifeless cloud, speeding its compression and giving birth to a new star.

The force of gravity pulls harder and harder as the particles of matter become closer and closer together, concentrating in the center of the swirling mass. The rotating disk of gas flattens as the centrifugal force pushes mass toward the edge of the disk. Gravity wins the battle as mass exponentially concentrates in the core of the disk. Boom! All of a sudden as internal pressures and heat (millions of degrees) overwhelm the center of the core, nuclear fusion erupts.

Internal heat and the ensuing collisions ionize the hydrogen atoms, stripping the nuclei of their solitary electron. The hydrogen and deuterium (one proton and one neutron) nuclei no longer bounce off one another in elastic collisions. They fuse together to form a helium nucleus in a tremendous burst of energy (nuclear fusion). Some amount of mass is lost in the fusion process and is converted to energy. A helium nucleus weighs approximately 99.3 percent of the sum of the four particles that comprise it: two protons and two neutrons. A small portion of the rest mass of the resulting helium nuclei is converted into a tremendous quantity of energy in the form of highly energized photons (the quanta of the electromagnetic force).

The newborn stellar mass performs some housecleaning in the dusty cloud of gas from which it was formed. Matter not captured by the burning sphere is ejected farther out in the disk. This excess material forms planets, planetoids and moons. Planetary moons are formed for the most part by excess material trapped in orbit by a forming planet. Candidate lunar material floating along between newly forming planets may also be captured by the gravity of a planetary body. The moons become natural satellites of the planetary bodies, as the planets became natural satellites of their mother sun. The surrounding space of the inner disk is swept clean by the planetary bodies. Remaining matter is exiled to a halo around the newly formed solar system in the form of asteroids and comets. They travel in very large elliptical orbits around the system's center of mass—the star. A few eventually make their way to the inner solar system for a round of target practice with an occasional, unlucky planet. Current extinction theories blame a cometary collision (some 65 million years ago) for the disappearance of the dinosaurs here on earth.

Planetary collisions by rogue comets and asteroids are commonplace, with hundreds of earth-crossing asteroids eyeing the earth. With limited funding, astronomers have mapped only a small percentage of these bodies. Difficult to detect, an asteroid on a collision course with the earth may give only a few days warning. The extinction level event that ended the reign of the dinosaurs on planet earth was caused by a cometary body about seven miles wide that crashed into the Gulf of Mexico near the Yucatan Peninsula. The energy of impact vaporized the comet on impact with the atmosphere, sea and sea floor. The blast sent a shockwave and ensuing firestorm circumnavigating the globe. Those species not instantly incinerated by the blast were doomed to die a horrible death in the following months. The impact and fires blew dust and smoke into the atmosphere blocking out the sun. The resulting worldwide winter killed off most of the plant life. Thus, the herbivores starved to death and eventually so did the carnivores. All of the dinosaurs were killed along with three-quarters of all species on earth. Following this planetary cleansing, mammals became the dominant species to roam the planet some tens of millions of years later. Like planetary collisions, the formation of the solar system was a violent process, and continues on at an abbreviated schedule even today.

STARDUST

We are truly created from the ancient remnants of stars—stardust. In fact, all matter—our bodies, our cars, our coffee cups and the matter that composes the ground beneath us—originated from the stars. "Ashes to ashes and dust to dust." The phrase is absolutely true. Our bodies, talking purely chemistry, are magnificent conglomerations of complex molecules. Taken one step further, these complex molecules are aggregations of simpler molecules. Molecules are made from common elements. The elements, or atoms, that constitute the matter in our bodies are common elements of the periodic chart—mostly hydrogen, carbon, nitrogen, oxygen and calcium. These elements are constructed in the cores of stars. Hydrogen is the most numerous element in our bodies, and is the most numerous within stars and in the universe by no coincidence.

The *Hindenburg*, one of the largest rigid airships ever built, ferried wealthy international passengers back and forth across the Atlantic in the mid 1930s. She was German by design and by registry. She demonstrated German superiority in design, engineering and elegance. However, the *Hindenburg* had a single detrimental flaw—she was filled with hydrogen. Hydrogen is the lightest element (granting the *Hindenburg* her loftiness), but it is also one of the most volatile, easily combustible elements in the universe.

Lakehurst, New Jersey saw her demise in May 1937. She was on final approach to her mooring mast when she suddenly and unexpectedly burst into flames a few hundred feet from the ground. The *Hindenburg* was destroyed when (most likely) a spark of static electricity ignited an envelope of its buoyant and highly combustible hydrogen gas.

The destruction of the great airship is an example of a chemical reaction. There are many more examples. Coal and natural gas combusted in huge chambers convert water into steam, which turns massive turbines leading to generators, and supplies the nation and the world with its energy needs. If a tanker truck full of gasoline detonates, a great deal of energy is released in the chemical reaction of its combustion. The reactions described above all release energy, but none are severe enough to create or destroy the chemical elements.

The hydrogen detonated in the *Hindenburg* disaster combined with oxygen in the air to form water vapor in an intense release of energy. However, only the chemical bonds were rearranged—only the outer electron shells were affected. Negatively charged oxygen ions are willing to give up two electrons to two positively charged hydrogen ions forming a water molecule, but the hydrogen atoms remain hydrogen and the oxygen atoms remain oxygen.

Similarly, chemical reactions convert the coal and natural gas in power plants to other compounds by sharing and rearranging outer electron shells, but the elemental components (the nuclei) remain unchanged. In the above examples, no nuclear reactions take place. Chemical compounds are converted from one to another simply by adding or releasing a little energy by rearranging the way in which the electrons in the valence shells are shared. Here, the nucleus of the atom is unaffected.

In nuclear reactions, such as those centered at the cores of stars, elements are created from the building blocks of others. These reactions involve the nucleus of the atom and release a tremendous quantity of energy due to the power of the nuclear strong force. Hydrogen is converted into helium and helium into carbon and on up to iron in the fusion reactions that take place within a star's core. All of these elements are spread throughout the galaxy when a star explodes. All of the elements on the earth were created from nuclear reactions deep within the core of a once vibrant, ancient star and perhaps many stars. These elements, blasted away from their mother star, found their way to the conglomeration of a hydrogen and helium gas cloud that formed the solar system long ago. These elements became incorporated into the planets (including earth) that now orbit their sun.

Before nature used biology as a tool to forge life forms on this once barren rock, she first needed matter with which to fill her cauldron. She turned to physics—nuclear physics to be exact—the physics of manipulating atomic nuclei. Nuclear reactions (natural nuclear reactions, not the ones trifled with by man) proliferate in the cores of stellar bodies—in stars not in planetary bodies. Planets the size, i.e. mass, of Jupiter and smaller have insufficient mass to smash atomic nuclei together. Although massive on a planetary scale, not even Jupiter has enough heat and pressure in her interior to bring two atomic nuclei close enough to overcome the electromagnetic force and initiate a nuclear fusion chain reaction.

All hydrogen that exists today was created moments after the Big Bang, shortly after the quarks and leptons precipitated out of the energy flux. In this period of time (several minutes) the energy density was many orders of magnitude more intense than the center of the sun. The infant universe was a nuclear furnace.

Some of the hydrogen captured neutrons to form deuterium. Shortly after, the deuterium (a hydrogen atom whose nucleus contains one proton and one neutron, "heavy hydrogen") fused with other deuterium to form helium (an atom whose nucleus contains two protons and two neutrons). In a cosmic blink of an eye approximately 23 percent of all matter (hydrogen) in the universe was converted to helium. A very small portion was converted to lithium (atomic number 3) via this early nuclear fusion process. This process of fusion to produce helium and traces of lithium continued for only a few minutes.

At this period of time in the fetal universe nuclear reactions slowed to almost a standstill. The universe was experiencing a period of extremely rapid inflation. As a body of gas expands it is cooled, and its energy density rapidly diminishes. The universe quenches its fire.

The universe is expanding. Like an expanding cloud of gas and dust from a catastrophic explosion the universe spreads out in all directions (of space and time). It is gravity's responsibility to bring things together. Without it, the expanding cloud of gas and dust would keep expanding forever, and the universe would become less and less dense. The average density of today's universe given the vast voids between neighboring galaxies is about 0.2 atoms per cubic meter. For comparison, the average density of water is 3×10^{28} atoms per cubic meter.

Our mature universe varies in density to the extremities of galactic cores of billion-solar-mass black holes to practically zero between the massive vacant voids between walls of galaxies. Gravity is the glue that binds interesting structures together. Gravity is the glue that sought out minuscule density fluctuations in the infant universe to clump matter together. These slight lumps of density in the early universe were the seeds of massive structures evident today—superclusters of galaxies. Without gravity, the universe would be a pretty unremarkable place—a perfectly dark void of a near vacuum, with an occasional hydrogen or helium atom for every five cubic meters of space—a size comparable to the inside of a car. If the force of gravity differed more or less by only a tiny fraction of its value, the universe would have contracted long ago into a Big Crunch, or (if too weak) galaxies and stars would never have been able to form. Gravity is a finely tuned parameter. Perhaps this gravitational constant value was prescribed by some higher means for a delicate machine?

Thanks to gravity, the universe is able to forge heavier elements. From the furnace of the Big Bang, we add hydrogen, helium and a little lithium to the bag of building blocks—pretty boring from the chemist's point of view. Without heavier elements, no complex molecules could ever form. The universe will have to wait for tiny density fluctuations in that early epoch to become the seeds of galaxies, quasars and galactic walls. Only then in the stellar furnaces (i.e., stars), are heavier, more complex elements formed.

Some several hundred million years later, stars are beginning to form. The tremendous supply of hydrogen congregated into enormous clouds, began to condense, and initiated the first stellar furnaces. These stars were composed entirely of hydrogen and helium (and that minuscule amount of lithium not worth mentioning). Thus begins the manufacturing of the heavier elements.

STELLAR FACTORIES

It is now up to stars to produce all of the elements more complex than hydrogen. There are 92 naturally occurring elements from hydrogen, atomic number one, containing one proton, all the way up to uranium, atomic number 92, containing 92 protons. Scientists, with the aid of particle accelerators and breeder reactors, have formed elements heavier than uranium, but these elements are unstable and eventually decay into simpler forms of matter (between atomic number one and 92). The insides of stars are breeding grounds for the building blocks of matter.

Our sun, for example, is a G-class main sequence star. It is a typical yellow star of average mass and average luminosity. It is on the fairer side of being middle-aged at approximately five billion years old, and will continue to burn hydrogen at a steady rate for another five billion years. The sun converts hydrogen into helium in its core. When the sun reaches the end of its life, it will have converted almost all of the hydrogen in its core to helium, resulting in a core of almost pure helium.

The constant fusion of hydrogen creates enough kinetic energy to produce a constant outward pressure. This continuous outward pressure balances the massive core against the constant threat of gravitational collapse. Fast-forwarding some five billion years into the sun's future, the hydrogen fuel is all but used up, so there is no more outward pressure. The core begins to collapse.

Helium nuclei, containing two protons and two neutrons, have twice the electric charge of their hydrogen counterparts. Therefore, they need to collide faster, with greater vigor, to overcome the electromagnetic force of repulsion between the like-charged nuclei. As the stellar core collapses further and further, it heats up hotter than temperatures and pressures required for hydrogen fusion. The core heats up, increasing the pressure until the helium nuclei can react. Poof! The initiation of helium has begun, and further collapse of the core is balanced by the kinetic energy of helium fusion. The core is in equilibrium again for a while.

The helium burning stage burns tremendously due to the increased temperatures and pressures in the core necessary for helium fusion. This increased energy output pushes against the sun's outer layers, and the sun expands into its red giant stage. The star's surface layers appear red due to the cooling of the plasma as it expands. The surface of the sun now reaches some 50 million miles into space, halfway to the earth. Provided that mankind survived this long, life would be almost impossible on the scorched earth, void of oceans and atmosphere.

This is where the stellar alchemy stage becomes interesting. A helium nucleus cannot simply fuse together with another. A nucleus with eight nucleons (protons and neutrons) is unstable. Two nuclei of helium fuse, and before it can decay another helium nucleus joins the unhappy couple, so in this union three's company. The additional helium nucleus fuses to the other two forming a stable carbon-12 nucleus—one of nature's most stable elements. The process takes tremendous temperatures and pressures to fuse the third helium nucleus before the beryllium nucleus decays. The research of helium fusion to form carbon is attributed to Edwin Salpeta.

Eventually the helium fuel is used up. When the helium is used up, the core contracts even further. The centers of heavy stars (much heavier than the sun) may reach temperatures of a billion degrees. The "helium flash" soon withers down and the star settles down as a white dwarf about the size of the earth at a temperature of 10,000K.

A white dwarf is a star in a curious predicament. Stars as massive as our sun will eventually be squeezed down to the size of the earth. After the helium fuel is consumed and converted to carbon and the fusion fires dwindle, there is insufficient outward pressure to support the core. The stellar core continues to collapse under the intense gravitational tides. The star contains insufficient mass to perpetuate the fusion reaction. As heavier and heavier elements are synthesized (now up to carbon—atomic number 6) more and more energy is required in the form of heat and pressure, and less energy is produced from the reaction to add heat to the fire. The stellar compression continues and is eventually halted to form a white dwarf.

No two particles can occupy the same space simultaneously. Only so many people can fit inside a room, although the owners of some bars and nightclubs during happy hour may think otherwise. Similarly, only a finite number of electrons, protons and neutrons can occupy a given volume of space. The gravitational collapse continues no more. Further collapse is prevented by a quality of nature discovered by Wolfgang Pauli, stated in the Pauli exclusion principle:

No two particles can occupy the same quantum state at the same time.

Subrahmanyan Chandrasekhar attributed the Pauli exclusion principle to white dwarfs in the 1930s.

We now have carbon to add to nature's caldron, and can now account for all elements up to atomic number 6, from hydrogen to carbon. The complexity and diversity does not stop here, but we require stellar factories much more massive than our own—our sun. In extremely massive stars, the fusion reaction does not cease with carbon. Carbon nuclei join with other helium nuclei to form oxygen (atomic number 8). Oxygen joins with another helium nucleus to form neon. With internal stellar kinetic energies becoming increasingly intense, the heavier nuclei fuse together to form silicon. Silicon nuclei join to form nickel, cobalt and iron.

Each higher reaction produces less energy until the process eventually reaches iron. Iron actually requires an input of energy for nuclear fusion to take place. This "iron curtain" prevents the fusion of more complex elements.

We now have all of the elements up to iron (atomic number 26) to add to nature's caldron.

The massive core of the collapsing giant left behind from the blast furnace that created iron is a perplexing entity. No further fusion can take place. Therefore, the core is no longer supported against the outer layers of the star. It becomes an incredibly dense place, compressed to a radius of about ten miles. A spoonful of matter weighs as much as a mountain here on earth. The surface is perfectly smooth, a mountain here is no more than a millimeter high. Gravity squashes everything completely flat. The star is a neutron star.

The electrons and protons that were supported from further compression in the white dwarf by the Pauli exclusion principle are compacted together. The electrons and protons in the star are squeezed together to form neutrons. Neutrinos are also produced in the reaction that escape unimpeded through the outer stellar layers and out into space at close to the speed of light:

$$e^- + p \quad \rightarrow \quad n + \nu_e$$

The star consists almost entirely of neutrons, with a thin iron crust.

Free neutrons, or solitary neutrons isolated from an atomic nucleus, are unnatural. A neutron by itself will decay in approximately ten minutes, but neutrons inside an atomic nucleus joined together with protons by the nuclear strong force do not decay (in a reasonable period of time). Similarly, the neutrons within a neutron star are stable. The entire neutron star acts as a single gigantic atomic nucleus (with a ten-mile radius). The Pauli exclusion principle prevents further collapse—no two neutrons can occupy the same quantum state at the same time.

Still heavier elements exist in nature, up to uranium (atomic number 92). Up to this point we can account for all elements up to iron (atomic number 26).

The rapid collapse of a star into a neutron star produces a cataclysmic blast. The star explodes in a brilliant flash called a supernova. For a brief time, the supernova outshines the entire galaxy. The stupendous blast sends a shock wave to the outer shell of the star instantly fusing heavy nuclei into still heavier nuclei. The process produces elements including platinum, gold, silver and lead all the way through the periodic table up to uranium. The outer shell of the star is blown off into space. The core settles back down to a neutron star.

We now have all of the elements up to uranium (atomic number 92) to complete nature's caldron.

Elements heavier than hydrogen are manufactured in stars, and are spread around by supernova explosions. Supernovae serve to seed the galaxy with the complex elements. All elements up to iron are created without the aid of supernovae, but what good do these elements provide if they are locked up in a massively dense stellar core? The earth contains traces of these heavy elements, but how did they get here?

The sun did not produce them; it never will, for it is not massive enough. Sometime in the distant past the interstellar cloud of gas that eventually formed our solar system was seeded with elements of increasing mass and complexity—from carbon to iron to gold to uranium. A distant supernova, perhaps several, sent these elements pouring into the simple interstellar cloud. A dying star's last duty after breathing life-energy into its solar system for billions of years is to spill its guts into interstellar space so that other stellar systems may use its raw material to build complex molecules and eventually planets and people. A dying star breathes life into a lifeless nebula. So, we truly are built from the ashes of a dying star. Each time matter is cycled and recycled. All protons (10^{78} of them) have existed since the instant before the nucleosynthesis stage, moments after the Big Bang. Each time matter is recycled through stars. We truly are created from stardust.

STARS: DWARFS AMONG THE GIANTS

The most fundamental property of a star is its mass. Mass dictates lifespan and course of evolution for a star. A star's temperature, i.e. color, and luminosity depend on its mass. The course of life the star will take depends solely on its mass. Will it burn brightly, quickly but furiously, and end its life after a few million years in a stupendous grand finale explosion—a supernova? Will it burn steadily for tens of billions of years as an average yellow star, eventually exhausting its hydrogen and then helium fuel to settle down as a white dwarf and then dimmer still as a red dwarf for perhaps a few trillion years? Will enough mass remain after a fiery death to collapse the core into nonexistence—a black hole? The star's mass decides its fate.

It is ironic, contrary to intuition, but the more massive a star, the shorter its life. Extremely massive stars live short, violent lives. Stars of average mass, like the sun, live a very long, uneventful existence. Gravity maintains a tighter grip on more massive stars. The more mass a star contains the stronger its gravitational force. The intense gravitational influence of an extremely massive star squeezes the stellar core tighter and tighter. The increased pressure at the star's center presses the hydrogen nuclei closer and closer. The greater pressure at the center of a massive star causes unusually high temperatures. The higher temperature causes the hydrogen fuel to react much more violently, thus consuming its seemingly abundant fuel supply very rapidly.

Therefore, more massive stars are caught in a higher gravitational field, so their stellar cores are pressed tighter and squeezed harder resulting in higher temperatures and more violent nuclear fusion reactions. The increased nuclear activity consumes the hydrogen fuel more rapidly than an average star, and as a result the massive star lives a shorter life.

The mass of a star defines its aggressive nature and predicts its life span. A star's color depends on its surface temperature, which is directly related to its mass and stage of life. The color of a star is directly related to its surface temperature. A star radiating at a given temperature will emit a distinct color. The color of a star (blue, white, yellow, orange, red) dictates its surface temperature. The Harvard classification of stars, organized in about 1900, arranges stars according to their type, color and temperature (Breithaupt, p73):

O	blue	30000K
B	blue-white	20000K
A	white	10000K
F	yellow-white	8000K
G	yellow	6000K
K	orange	4000K
M	red	3000K

The sun is a class-G type yellow star with a surface temperature of about 5800K.

The brightness or luminosity of a star depends on its surface area, or size. A red star like the class-M5 red dwarf Proxima Centauri is very close to us but very dim since it is so small. The red supergiant Betelgeuse is very bright due to its immense surface area—its diameter is as large as the orbit of Mars! Both the red dwarf and the red supergiant are about the same temperature at the surface (3000K), but the giant is orders of magnitude brighter. The greater the surface area, the more light emitted, and hence the brighter the star.

Stars may be classified by their spectral class (color) and their absolute magnitude (brightness). Such a classification, the Hertzsprung-Russell diagram, was developed independently by engineer Ejnar Hertzsprung and Princeton astrophysicist Henry Norris Russell around 1914. Using this method, a star's absolute magnitude (brightness) is plotted on a graph versus its spectral content (color). The resulting chart places a majority of stars along a thin band from the lower-right corner (cool, dim stars) to the upper-left corner (hot, bright stars). This thin band represents the main sequence stars, of which the sun is a member. This band of main sequence stars constitutes the bulk of the stars in the galaxy.

Main sequence stars burn hydrogen in their cores converting about ten percent of it into helium. Giant stars lie in the upper-right portion of the Hertzsprung-Russell diagram, and dwarf stars lie in the lower-left.

Starlight reveals more than just temperature. Starlight also announces to all observers what a star is made of. The elemental composition of a star may be determined by observing the light emanating from its surface. It is utterly amazing!

The substance of a luminous beast is known without even touching it or taking a small sample of it. Starlight may be decomposed by passing it through a thin slit and then through a prism and projecting the resulting band of light onto a flat surface. The light is separated into constituent wavelengths by the prism, just as tiny droplets of water separate sunlight into the colors of the rainbow.

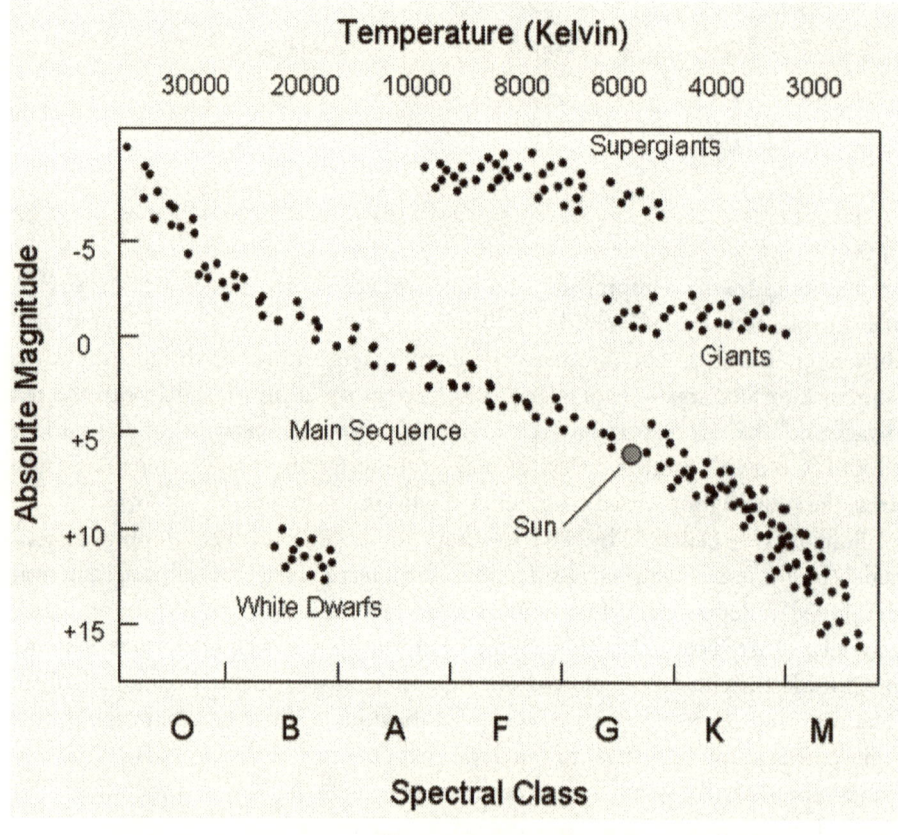

The Hertzsprung-Russell diagram relates the spectral class (color/temperature) of a star with its absolute magnitude (brightness level). A majority of the stars in the Milky Way——sun included——belong to a thin band that bisects the diagram from upper-left to lower-right. The stars within this band are known as the main sequence stars.

Certain chemical elements absorb light at very distinct wavelengths. An electron that absorbs a photon becomes excited and moves to a higher energy state, but that electron will absorb a photon at a specific wavelength. The outer electrons of individual elements absorb particles of light—each element at a distinct frequency. Some of the light produced in the outer layers of a star is absorbed by the elements in the surface of the star.

Sodium, for example, in the sun's atmosphere, absorbs a certain distinct wavelength of the light emanating from the surface. When this light is decomposed (through a prism) a thin dark line will appear in the exact location where the wavelength of light was absorbed. A continuous spectrum (like a rainbow) should appear, but since certain distinct frequencies are absorbed, thin dark lines (absorption lines) appear across the band of decomposed light. By analyzing where these predictable bands appear, astronomers can predict very accurately what elements exist in the star's outer layers.

Starlight provides yet another extremely beneficial utility. By analyzing how far these dark bands in the starlight's spectrum are shifted from their predicted wavelengths, astronomers can predict the star's speed, relative to the earth, and even its direction. If the dark bands in the starlight's spectrum are shifted toward longer wavelengths (toward the red end of the rainbow–redshifted) the star is moving away from us. The amount of shift is directly proportional to its velocity. If the dark bands are shifted toward shorter wavelengths (toward the blue end of the rainbow–blueshifted) the star is moving toward us. Hubble used this anomaly in the light emanating from galaxies and found that almost all distant galaxies are redshifted and hence moving away from us. Seeing is believing—light alone can reveal many secrets.

Stellar bodies exist in unbelievable variety from tiny dwarf stars the size of small planets to gigantic supergiants whose diameter would encompass the entire orbit of Saturn. Stars continue to be born and inevitably die. Some die a slow, cold and quiet death, while other more massive stars end their lives in cataclysmic explosions spewing their innards across the galaxy, seeding young stellar nurseries with heavy elements. Some in their dying moments leave a core so dense that it implodes, winking out of existence and taking everything nearby with it, including particles of light.

Dwarf stars are the smallest of the main sequence stars. Dwarf stars are tiny—the size of small planets—and they are the most populous stars in the galaxy. The closest star to the sun—Proxima Centauri (actually the third member of a close-knit trinary system)—is a red dwarf, but is too dim to be seen with the naked eye.

Brown dwarfs are objects containing only a fraction of the sun's mass but many times more excessive than the massive gas-giant planets of our outer solar system. Brown dwarfs are failed stars—stars with insufficient mass to initiate nuclear fusion. Brown dwarfs emit no visible light, but emanate small amounts of infrared radiation due to internal pressures. Temperatures at the surface of a brown dwarf are more comparable with a planet rather than a star.

White dwarfs are stars similar in mass to the sun but similar in size to the earth. They appear in the lower-left region of the Hertzsprung-Russell diagram. White dwarfs are hot—evident from their white color—but too dim to be seen without the aid of a telescope. Stellar matter from a white dwarf is very dense—its entire mass has been crushed to the volume of a planet. A spoonful of matter would weigh as much as an automobile. White dwarfs are prevented from collapsing further by a quality of nature that states that electrons cannot occupy the same state simultaneously. In other words, electrons can be squeezed together only so far. This is the Pauli exclusion principle. White dwarfs are the end result of a red giant that has run out of its helium fuel. They are actually the tiny, exposed, helium-rich cores of a red giant. White dwarf stars are the grave sites of red giants. The sun will become a white dwarf following its red giant phase in some five billion more years.

A white dwarf is a star as massive as the sun but squeezed down to the size of a planet, some ten thousand miles or so across, and the Pauli exclusion principle sustains it from further collapse. Neutron stars break the mold of the typical star. Neutron stars are many times more massive than the sun, but exist in a spherical volume some ten miles or so in diameter. The extreme mass and immense density of a neutron star overwhelm the Pauli exclusion principle and cause the electrons to crash into the protons of the stellar body, squeezing them together to form neutrons. Momentarily the stellar body is transformed almost purely into neutrons. The entire neutron star exists as a single, but gigantic, atomic nucleus the size of Manhattan. Neutron stars are extremely dense—a spoonful of material would weigh as much as a mountain. For this reason the surface is perfectly smooth, and Mount Everest would be no higher than a grain of sand. Neutron stars are prevented from further collapse by the neutrons' obedience to the Pauli exclusion principle.

A pulsar is a spinning neutron star whose intense magnetic field is inclined with its axis of rotation. As the pulsar spins, its magnetic field is swept across space in a beam of radio waves like some sort of cosmic lighthouse. When the beam sweeps across the plane of the earth, astronomers see it as an intense pulse of regularly occurring radio waves. Their warbling axes sling radio waves out into space like a high-pressure water hose slung around and around a child's head. Pulsars can rotate as rapidly as a thousand times per second (millisecond pulsars). Here a solar day lasts only a millisecond (a thousandth of a second).

White dwarf and neutrons stars are a cosmic curiosity due to their minuscule size and immense density. The giants (red giants and supergiants) are equally impressive due to their behemoth proportions. When a star such as the sun exhausts its hydrogen fuel at its center, its core begins to contract. Gravity wins the constant tug-of-war as the hydrogen-to-helium fusion fires dwindle. As the core cools, it no longer maintains a constant outward pressure to support itself against the crushing force of gravity, so the core collapses and the outer layers expand. As the core contracts it begins to heat up as any other thermodynamic reaction with decreasing volume. Helium requires more heat and pressure to initiate nuclear fusion than hydrogen (since helium contains twice the repulsive force of hydrogen), but the increasing heat and pressure triggers helium fusion as the helium-rich core of the sun-like star collapses. Helium fusion reactions burn much more violently than hydrogen fusion. Further collapse of the core is averted by the rise in kinetic energy sparked by the helium reaction, but the increased energy output of the core's nuclear reactor and the change in gravitational tides push against the star's outer layers. The outer layers expand and the surface turns a deepening red as the concentration of radiated energy disperses and the surface cools.

When the sun exhausts its hydrogen fuel at its core in about five billion years, and begins to burn helium during its red giant phase, its photosphere (outer layer) will reach past the orbit of Venus, halfway to the earth. Some red giants (more massive than the sun) boast diameters larger than the earth's orbit around the sun. Antares, for example, is a K-class supergiant in the southern constellation of Scorpius whose diameter is large enough to overtake the orbit of Mars. Red giants exist in the upper-right portion of the Hertzsprung-Russell diagram, and supergiants dwell in the upper region. Some supergiants are so immense that their photospheres would stretch out to the orbit of Jupiter and even Saturn.

Depending on their mass, stars may exist in a variety of exotic forms, but their deaths can momentarily overwhelm their entire parent galaxy. Stars similar in mass to the sun will undergo a quiet, uneventful and slow death. After extinguishing its supply of hydrogen fuel at its center, its core will contract and its outer layers will expand. As the core is compressed, the helium is ignited while its outer layer—its photosphere—expands and glows a deepening red as it cools. After the sun endures its violent short-lived red giant stage, it will have blown off most of its outer layers. Its naked helium-rich core will remain as a white dwarf.

In its final stage, the sun will live out its retirement as a tiny white dwarf for perhaps hundreds of billions of years (if left undisturbed and assuming the universe does not collapse).

Stellar bodies more massive than the sun (at least 1.44 times as massive—the Chandrasekhar limit) undergo extremely violent deaths—episodes so violent that for a short time they outshine the entire galaxy in which they inhabit. The term supernova is derived from the Latin root *nova,* which means "new star." That is what the ancients saw. In the several thousand years of civilization (before the advent of technology) the few supernovae that have been witnessed have appeared as a new star, never before seen.

A Type Ia supernova occurs from within a binary system. A Type Ia supernova results from the explosion from a member of a binary system—a white dwarf. A dwarf star waltzing in close unison with a larger, less dense companion siphons part of the larger beast's photosphere. The dwarf steadily gains mass from its larger stellar partner. While on a sailing voyage from Bombay, Subrahmanyan Chandrasekhar, the Indian astrophysicist, calculated the amount of mass required for a dwarf star to collapse. This amount of mass is 1.44 times the mass of the sun and is known as the Chandrasekhar limit. As stated previously, white dwarfs are prevented from further collapse by the Pauli exclusion principle that no two electrons can occupy the same quantum energy state at the same time. A dwarf star in a binary system that steadily sucks matter from its less dense partner slowly gains mass. Once the Chandrasekhar limit is reached, the pregnant dwarf star can no longer contain its inevitable collapse. The dwarf collapses further and billows forth a stupendous explosion—a supernova.

Type Ia supernovae have very predictable brightness levels. This fact makes them beacons to cosmic surveyors—astronomers wishing to estimate the distance to a galaxy with such a beacon. As viewed from earth, the brightness level of a Type Ia supernova is proportional to its distance. Hence, the distance to a far away galaxy with a Type Ia supernova eruption can be measured with amazing precision. A Type Ia supernova's brightness and frequency of occurrence are predictable. A given Type Ia supernova will erupt at regular intervals when enough matter has been siphoned from its stellar dance partner.

A Type II supernova results from the transition of a star to a neutron star. A supernova marks the end of a giant. This formidable event will, for a brief time, outshine its entire parent galaxy. When a red giant or supergiant has consumed its nuclear fuel—fusing helium into carbon, oxygen, neon and silicon on up to iron—its core collapses beyond the density of a white dwarf to form a neutron star.

The process of squeezing the electrons into the complement of nuclei within the core results in a blindly rapid, stupendous blast. The supernova produced by the collapse of a white dwarf to form a neutron star releases most of its energy into neutrinos. The shockwave produced by the supernova moves outward through the giant's outer layers fusing the constituent matter into elements heavier than iron—platinum, gold, silver and lead up to uranium. The blast serves not only to synthesize the elements heavier than iron, but it also expels the matter into interstellar space. Distant clouds of hydrogen and helium gas such as the one that formed the solar system five billion years ago are seeded with heavy elements this way. A Type II supernova leaves behind in its wake a neutron star.

Perhaps the strangest of all interstellar phenomena—a black hole—results from the collapse of an extremely dense body. The continuum around the dense body becomes so warped (as general relativity formulates) that the body actually "winks" out of normal space. A black hole is produced from an extremely intense gravitational field, so strong that nothing, not even light, can escape its persistent grasp.

In order to understand the existence of a black hole, an entity whose gravitational tides are so immense that not even light can escape, one must appreciate the concept of escape velocity. Escape velocity is the velocity required to overcome the influence of a massive body's gravitational field. Escape velocity depends only on the mass and size of the body producing the gravitational field (such as the earth), and is independent of the minuscule mass of the escaping object (such as a rocket, baseball or particle of light).

Satellites boosted into orbit by powerful rocket engines attain orbital velocity (about 17,000 miles per hour for the earth), but never exceed the earth's escape velocity, hence they remain in orbit. Escape velocity for the earth is about 25,000 miles per hour. This value is independent of the mass of the traveling object like the satellite. If an extremely strong baseball pitcher (much stronger than this author!) was able to throw a baseball faster and faster, it would obviously travel farther and farther. If he could somehow, without breaking his arm, attain a speed of 17,000 miles per hour, the baseball (neglecting atmospheric drag) would achieve orbit around the earth. If the pitcher could throw at a speed of 25,000 miles per hour, the baseball would leave earth's orbit and travel in orbit around the sun. To leave the sun's gravitational influence, the object must travel even faster than this.

Speed is everything. To escape the gravity well of the earth, a spacecraft, such as the *Voyager 2* that explored Jupiter, Saturn, Uranus and Neptune, must achieve a velocity faster than 25,000 miles per hour, else it will remain in orbit.

The Apollo spacecraft that landed men on the moon achieved a speed greater than 25,000 miles per hour to escape earth orbit for a lunar trajectory.

Escape velocity is a function of an object's (star or planet) mass and radius. If the radius of a star or planet was reduced, keeping its mass constant, its gravitational field would become more concentrated. Newton's law of gravity depends on distance (in this case radius). Reducing the radius of the object has concentrated the force of gravity. Objects on the surface of the body now seem heavier. The escape velocity from its now reduced surface must increase to overcome the increased force of gravity.

A collapsing star that has been compressed by gravitational tides boasts a greater escape velocity than when the star was larger in size. Remember the speed of light, although tremendous, is finite. An object whose size has been compressed enough for its escape velocity to exceed 186,000 miles per second (the speed of light) will become a black hole. The object is termed "black" because no photonic information—no light—can leave its radius of influence—its event horizon.

The radius of the object required for the escape velocity to equal the speed of light is called the Schwarzschild radius. Karl Schwarzschild was a German astronomer who contemplated such unworldly interstellar entities during the gruesome turmoil of World War I while serving as an artillery lieutenant on the Russian front. Schwarzschild found that the more massive an object, the larger the Schwarzschild radius. The sun would become a black hole if it were squeezed by some cosmic press to a radius less than two miles. The earth would achieve black hole status if it could be pinched down to the diameter of a quarter in your pocket.

A black hole is "black" because light cannot escape its surface. Since nothing can travel faster than light (as per the special theory of relativity), anything pulled into a black hole becomes imprisoned. The boundary around a black hole that cuts itself off from the rest of the universe is called the event horizon. It is a boundary—a point of no return for any matter that is pulled in. The diameter of a black hole depends solely upon the object's mass. A more massive object will exhibit a larger event horizon.

A black hole is formed naturally from the collapse of a massive dying star. When a star exhausts its nuclear fuel its core begins to collapse. The diminishing energy from the stellar core's nuclear engine cannot maintain enough outward pressure to counteract the inward pull of gravity, so the already compact core collapses further. The collapsing giant star sheds its outer layers in a stupendous supernova explosion and the core shrinks further. A star whose mass is greater than the Chandrasekhar limit (1.44 times the mass of the sun) will continue to shrink, squeezing electrons into atomic nuclei and resulting in a sea of neutrons: a neutron star. Stars with much greater mass will continue to shrink further past the neutron star phase, denser and denser until winking out of normal space and into a black hole. Mounting evidence supports the theory that a supermassive black hole, millions of times more massive than the sun, exists at the heart of the Milky Way Galaxy and acts as the engine that churns the galaxy. The gravitational pull of a black hole will seal the fate of any matter unfortunate enough to be caught in its wake.

Stars are bodies of mass all originating from some form of interstellar gas cloud composed mainly of hydrogen and helium and a sprinkling of heavier elements. The initial mass of the stellar body will dictate its course of evolution and longevity. Its mass will either preserve its life as a dwarf star or sign its death warrant as a supernova and perhaps a neutron star or black hole. Stars may be blindly energetic, stupendous volumes of gas that would encompass an entire solar system, or they may be tiny smoldering cinders the size of small planets. Stars may be ordinary.

The closest star to the earth, the sun, is a middle-aged, class-G type, yellow, main sequence star with an average surface temperature of 5800K. The sun first billowed life about 4.5 billion years ago and will continue to burn steadily for about 5 billion or so years more. Even though it breathed life into this once life-less rock, the sun is not distinguished or precious. It lies near the center of the Hertzsprung-Russell diagram, which classifies stellar bodies according to color (temperature) and brightness. The sun is very ordinary.

LARGE-SCALE STRUCTURE OF THE UNIVERSE

On a warm late summer evening peering through binoculars into the northwestern sky you can see the extremely faint tiny wisp of the great spiral, the Andromeda Galaxy. Although barely visible, it is a tremendous spectacle. You cannot focus on it directly, for it will quickly fade from sight. You must move your eyes back and forth on either side of the minuscule spiral smudge to keep it visible. Long ago, ancient light left this tiny galaxy so far away. It is difficult to fathom the time and space traversed.

In order to appreciate the beauty in such a seemingly insignificant speck, one must become aware of the great distances and large scales involved. That tiny wisp is the galactic home to about one hundred billion stars, like our own Milky Way Galaxy some two million light-years distant. That smudge is the sum of a hundred billion pinpoints of starlight, like the beach, which is the sum of hundreds of billions of tiny grains of sand.

Starlight that distant and compact is difficult to separate individually. Peering through the looking glass, one can imagine a similar being on a similar planet orbiting a similar sun on the outskirts of this distant island of stars looking back toward our direction and wondering about us. To this being, our galaxy is a tiny spiral smudge through his looking glass also. Perhaps there are distant beings looking back in our direction similarly contemplating their existence and the possibility of life elsewhere.

Galaxies are islands of stars. Beautifully analogous to atoms, galaxies contain mostly empty space just like atoms. An atom contains a central dense nucleus with distant (compared to their size) electrons orbiting around. A galaxy contains a dense central nucleus (perhaps a multimillion solar mass black hole) with its long spiral arms rotating around. Galaxies, although constituting billions and perhaps hundreds of billions of stars, are composed of mostly empty space. Stars are very far apart. Atoms, also, are mainly empty space. If the nucleus of an atom were the size of a Florida orange situated at the center of the earth, its neighboring electrons would be in orbit. Atoms are the smallest constituents of a substance. Further division of an aluminum atom, for example, into its constituent protons, neutrons and electrons would strip the aluminum atom of its title "aluminum." Similarly, galaxies are the smallest constituents of the large-scale universe. Hubble described galaxies as the basic unit of distribution in the universe—the basic building block of large-scale structure. Galaxies are the atoms of the universe, the building blocks of the universe. Looking deep into the distance, far, far away, individual galaxies become the grains of sand on the beach that is the universe.

The structure of the large-scale universe is both homogeneous and isotropic. This is stated as the cosmological principle. The universe, on the large scale is homogeneous. Matter (galaxies and clusters of galaxies) is evenly distributed. The large-scale universe is isotropic in that it looks the same in all directions. The cosmic background radiation emanates from every direction in outer space at virtually the same temperature (2.7 Kelvin, or 2.7 degrees above absolute zero). This temperature is extremely frigid but, nonetheless, above zero. It is a constant feeble reminder of a once violent past. The fact that space (on the average) radiates at precisely the same temperature all over the sky, to a few thousandths of a degree, is a clue indicating that the early universe reached a point of thermal equilibrium. For example, if a high-pressure volume of gas at a given temperature was expanded to a huge volume, its temperature would decrease (according to Charles' law), but every portion of the gas would agree to the same lower temperature. Every part of the gas would settle to the same temperature. The early universe existed as a high-density, high-pressure, high-temperature volume of energetic particles. As the universe expanded, it cooled, but cooled at the same rate, thus maintaining thermal equilibrium. It makes sense that the large-scale universe of today is both homogeneous and isotropic.

The universe is not perfectly homogeneous. This fact is a blessing, for it if were, there would be no galaxies. There would be no solar systems, no stars, no planets or people. These objects (galaxies, stars, planets and people) are much denser than the average density of the universe. If the universe was perfectly homogeneous throughout its current expanse of spacetime, the universe would exist simply as a vast cloud of hydrogen and helium with a universal density of 0.2 atoms per cubic meter. On the same line of reasoning the universe is not perfectly isotropic. If it were, every galaxy would be the same and equidistant from every other galaxy. There would be no higher structure to the universe.

To study large-scale structure is to understand cosmic history. The Big Bang model suggests that the vast superstructures apparent today (clusters and superclusters of galaxies) resulted from tiny density fluctuations in the very early universe. The tiny fluctuations evident today in the cosmic background radiation are proof that the early homogeneous universe was not perfectly uniform, and from these slight variations resulted the seeds of galaxies and superclusters of galaxies. This amount of granularity made galaxies possible. Local gravitational effects in these slight variations in the matter field became increasingly stronger and pulled more surrounding matter in. Matter clumped together to form mammoth frameworks. The huge structures are monuments of an ancient history, and to understand these formations would open doors to understanding how they were formed.

Margaret Geller of Harvard is one such stellar cartographer. She makes maps of galactic structure. Classifying the structure of superclusters is similar to categorizing clouds—there is no straightforward distinction from one class to another.

The structure of the large-scale universe begins with galaxies—the atoms of structure, the building blocks of the universe-at-large. Further classification continues up the grand scale to groups of galaxies, clusters, clouds, superclusters and finally to the most behemoth formations in nature—supercluster complexes of galaxies.

Groups of galaxies are typically a few million light-years wide. Groups are composed of a few large galaxies plus several smaller satellite galaxies. The Milky Way is a member of the Local Group. The Local Group is approximately three million light-years across. Two major galaxies dominate the Local Group: our own Milky Way Galaxy and the majestic spiral of the Andromeda Galaxy (M31). The Milky Way, containing approximately a hundred billion stars, is about one hundred thousand light-years in diameter. The Andromeda Galaxy is almost twice the size of the Milky Way.

We are certain, with reasonable assurance, that our Milky Way is a spiral. We cannot be absolutely confident, as we live inside and are looking out from our little porthole to the universe. Trying to visualize our galaxy from within her is like trying to give yourself a haircut with no mirror—you can only do so well! However, mankind has done very well with observation and deduction. We can be reasonably sure that our galaxy exhibits a spiral formation similar to her sister galaxy Andromeda. This line of reasoning follows from the observational fact that on a clear, late Arizona night the spiral arm of the Milky Way (the Great Sagittarius Star Cloud, consisting of many millions of stars some thirty thousand light-years from the sun) stretches overhead from north to south. This hazy, milky could extends in a sort of thick line across the sky, instead of being wallpapered all over the sky. Hence, we are viewing the galaxy edge-on, and further, the galaxy must form a sort of pancake formation. The distance the spiral arm stretches can be measured with reasonable accuracy to ultimately calculate the diameter of the Milky Way at about a hundred thousand light-years, just like the Andromeda galaxy. The distance to nearby stars may be measured by the parallax method and more distant stars by their redshift. The direction of rotation may also be calculated by the redshift of the receding sector and by the blueshift of the advancing sector. The Andromeda Galaxy rotates in the opposite direction of the Milky Way, suggesting that they were formed about the same time from adjacent storm clouds of hydrogen and helium gas.

The Milky Way Galaxy is about a hundred thousand light-years across; Andromeda, about 150 thousand light-years across. Both are just over two million light-years apart. This makes the two great spirals twenty times as far away as they are wide. In other words, the universe is a tremendously vast expanse, containing mainly empty space.

The Local Group boasts a third spiral galaxy, the Triangulum Galaxy (M33). Although significantly smaller and more diffuse, Triangulum looms large in the skies of the alien worlds in the Andromeda Galaxy. She is only seven hundred thousand light-years distant from Andromeda. From the skies of Triangulum, Andromeda must take up half of the sky! Triangulum's sparse nucleus and thinly populated spiral arms are home to approximately twenty billion stars (about one-fifth as populous as the Milky Way).

Besides the two majestic spirals that dominate, the Local Group contains a dozen or so smaller satellite galaxies. Two of the most notable and closest to our home galaxy are the Large Magellanic Cloud and the Small Magellanic Cloud. These galaxies were named after the explorer Ferdinand Magellan who circumnavigated the earth by ship. His crew saw these soft, glowing clouds in the night sky as they sailed into the southern latitudes. They are seen best from the Southern Hemisphere, and cannot be seen from the continental United States. These two satellite galaxies circle the Milky Way Galaxy in a tight orbit. The Magellanic Clouds were first thought to be nebulae within the Milky Way. The discovery of Cepheid variable stars within the nearby galaxies enabled a more accurate calculation of their distance, and put them well outside our galaxy. The larger of the two contains about fifteen billion stars (the Milky Way contains about one hundred billion stars) and orbits our home galaxy at a distance of 150 thousand light-years. The Small Magellanic Cloud contains about five billion stars and orbits at a distance of approximately 250 thousand light-years.

The next rung in the ladder of large-scale structure leads us to clusters of galaxies. Clusters are dense conglomerations of galaxies stretching some tens of millions of light-years wide. Clusters embody great ensembles of hundreds to thousands of galaxies. The most densely packed aggregate of galaxies close to us, in earth's sky is the Virgo Cluster.

The Virgo Cluster dominates our intergalactic neighborhood with some two thousand member galaxies. It embodies the center of the local supercluster. By its gravitational attraction the Virgo Cluster influences all local galaxies and groups, slowing down the local velocity field. The Virgo Cluster is massive enough and close enough to our Local Group (of which the Milky Way is a member) so that local galaxies are moving in the direction of the Milky Way instead of receding away with the rest of the universe. These galaxies headed our way exhibit a blue-shift in their spectrum. The galactic members of the Virgo Cluster are gravitationally bound to one another. The mutual attraction within the cluster prevents the member galaxies from receding from each other. Clusters ride along with cosmological expansion (predicted by Einstein, but discovered by Hubble) becoming farther and farther apart. Individual clusters themselves remain intact—neighbors that remain close friends.

Clusters are the largest gravitationally bound structures in the universe. The universe is expanding and spreading everything out in the process according to the Hubble constant (the rate of expansion of the universe). Clusters of galaxies, however, are gravitationally bound together and are prevented from expanding with the local velocity field. Local expansion is not evident in clusters due to the gravitational attraction of neighboring galaxies. The mutual gravitational attraction of the matter within a cluster suppresses the urge to grow with the expansion of the universe. Clusters become more distant from other clusters as a result of Hubble expansion, but the clusters themselves remain intact. Similarly, groups of galaxies also perform their waltz unaffected by the expansion. In fact, in our Local Group, the Milky Way Galaxy and the Andromeda Galaxy are moving toward each other at about fifty miles per second. They will eventually collide in some eight to ten billion years and probably merge into one galaxy. Fret not, for the universe will be almost twice its current age by then and the sun will be a smoldering cinder. Humans, if we have survived (and in whatever form), would have migrated to another star system by then.

Clusters of galaxies seem to congregate together more frequently than individual galaxies alone. Groups, large conglomerations and clusters of galaxies seem to be more evident than individual galaxies strewn here and there. Strings of linked clusters form extraordinary filaments throughout the universe and lead to superclusters spanning hundreds of millions of light-years (a billion trillion miles).

The universe is granular at smaller scales—scales of galaxies and groups of galaxies. From a distance of a few hundred thousand light-years, galaxies are incredible conglomerations of stars and star systems. Galaxies, although seemingly dense, are mainly empty space compared with their volume. Galaxies and their groups are granular when stacked up against clouds and supercluster complexes of galaxies. Galaxies are the grains of sand on the beach—the building blocks of the large-scale universe.

Supercluster complexes are perhaps the largest structures in the observable universe. The universe at this scale is similar to foam—like the foam that washes up on the beach as a result of the pounding surf. Vast voids of galaxies—some voids hundreds of millions of light-years across—are encased in walls of super-clusters of galaxies. These walls of clusters and superclusters of galaxies, and the voids they encompass, form a structure similar to soapsuds in a bathtub. The walls are great barriers to the tremendously vast, lonely voids separating tremendous populations of galactic clusters from unbelievable desolate darkness.

There is a higher order of structure to the universe, higher than simple conglomerations of enormous galaxies. Galaxies are not sprinkled evenly throughout space but, rather, are spread out into enormous filaments. This appears to be a common feature of the large-scale universe billions of light-years distant.

Sitting in a bathtub full of soapsuds you will notice millions of tiny, white bubbles. The bubbles appear evenly distributed—homogeneous. If you sink down, becoming completely encased and surrounded in the white foam of the bubble bath, you would notice that it looks the same in any direction you looked—isotropic. (This mild bath will not burn your eyes.) The structure of the foam is such that thin filaments of soap film conglomerate together, encompassing pockets of air, and forms millions of bubbles. On a scale of hundreds of millions of light-years, strings of linked clusters of galaxies form enormous walls that encompass vast, desolate voids. On a grand scale, these tremendous voids are the bubbles; the walls of linked superclusters of galaxies are the filaments of soap film that embody the soapsuds of the observable large-scale universe.

The universe is extremely complicated, yet remarkably simplistic. Perhaps there will remain more to universal structure than mankind may ever realize. Perhaps our universe may exist as a tiny bubble in a more unimaginably vast—if one could fathom—"multiverse." Given this extremely speculative conjecture, perhaps our universal laws of physics are merely a subset of a more complex set of laws. Perhaps our universe and the laws governing it are merely a fingerprint of a greater body. Mankind may never know.

PRIMORDIAL ENTITY

The universe is a fantastically complicated piece of machinery. Only a truly remarkable event could have transpired to manufacture all matter that exists in every configurable exotic form. The universe was formed from nothing. Well, perhaps not exactly "nothing" in the common sense. Perhaps the colossal fireball (which hardly serves this event justice) that created the universe was caused from the collapse of a previous universe into infinite density at an infinitesimal point. No one can be sure. It may be safe to say that no one will ever be sure of this sequence of events, for no information from the space and time of some previous universe could be passed to our spacetime continuum.

At the instant of the Big Bang the entire universe existed in its most simple form—energy. This infinite energy density existed in a point smaller than a proton. It had zero dimensions. Quantum theory allows particles to be created out of energy as particle/antiparticle pairs. The pair is referred to as a virtual pair, as the particles are seemingly created from nothing. For the most part the pair is usually annihilated in mutual destruction as a result of the violent nature of matter/antimatter, and their energy is relinquished to the vacuum of space as high-energy radiation (gamma rays). Nonetheless, particles may be created out of energy. One may ask, "Where did the energy come from?" In order to understand this concept, one must appreciate the nature of energy in the universe.

The total energy content of the universe is precisely zero. Matter manifests itself as positive energy. Matter is created (or transformed) from energy. All matter is attracted to all other forms of matter by the force of gravity. Two particles of matter at a given distance contain less energy than two particles separated by a greater distance because energy has been expended in the separation of these two distant particles. More energy has been spent by separating the farthest particles the greater distance. Thus, the gravitational potential between the two particles can be described as negative energy. Since the universe, at a very early age, was contained in a much lesser volume than the immense volume of today, the distribution of matter in the universe—positive energy—is approximately uniform in space. The negative gravitational energy exactly cancels the positive energy inherent in matter. Therefore, the total energy of the universe equals zero, and the law of conservation of energy is preserved.

The law of conservation of energy is not violated under the inflationary theory. Stephen Hawking explains that during the inflationary epoch, the energy density of the young, tiny universe remains constant while the universe expands. When the universe increases in size, the negative energy content increases proportionally with the increase of positive matter energy, so the total energy remains zero. All of the negative gravitational potential summed with all of the positive matter potential equals zero. As Alan Guth, the founder of inflationary theory, once put it, when explaining how the entire universe could be created from nothing, "It is said that there's no such thing as a free lunch. But the universe is the ultimate free lunch."

At the moment of creation the spacetime continuum—the entire universe—had zero dimensions. For in this instant spacetime unfurled. It is inappropriate to say that energy has "weight." Perhaps it's a little more appropriate to say that it has mass. As Einstein pointed out in his special theory of relativity, energy is mass, just in another form.

Energy and mass (matter) are different manifestations of the same entity. This relation is expressed harmoniously by $E = mc^2$.

The speed of light in a vacuum, c, is a very large number (3×10^8 m/s or 186,000 miles/s). This number multiplied by itself is a huge number, and multiplied by a speck of mass turns into a monstrosity of energy. With this destructive (if man so chooses) amount of energy, nuclear weapons with unthinkable devastation are possible. Without this power the sun in the sky and the stars at night would not shine.

A minuscule amount of matter may be transformed into a tremendous quantity of energy. Less than one percent of the sun's mass is converted into energy when hydrogen nuclei are fused into helium in the core of the sun.

With this understanding of matter and energy, it is not inconceivable that the entire universe fits into a size smaller than a proton. Energy has no dimension, so it can become extremely compact. Zero mass and infinite energy—these are the conditions that existed at the instant of creation.

Spacetime then unfurled. Three spatial dimensions and one temporal dimension began to expand leaving the other higher dimensions infinitesimally curled in on themselves. For a moment in time immediately following the Planck epoch when the universe was about 10^{-34} seconds old, the radius of the universe doubled in size in equal increments of time. Twentieth century physics cannot describe the conditions that existed this early in the universe. The laws of nature discovered and described so far by science break down in the extreme conditions, in the unimaginable energy and quantum fluctuations that existed in this early universe. Mathematics tends to break down when dealing with infinities.

INFLATION

The standard model of the Big Bang theory has developed some discrepancies. To understand these we must go back in time to the primordial universe to the moment when spacetime unfurled and matter and antimatter were created. This moment of time took place immediately following the Planck epoch when the universe was about 10^{-35} to 10^{-32} seconds old.

First of all, the universe is larger than the distance light could have traveled since the creation of the universe. This is referred to as the *horizon problem*. Einstein's special theory of relativity states that nothing can travel faster than light. Light is the cosmic speed limit for anything traveling in the cosmos. Then how can the universe be larger than the distance light could have traveled in about ten billion years since its creation? What is the horizon? Just as the horizon where the sea meets the sky is the limited to the distance we can see from a beach, the horizon of the universe is also the boundary from which we cannot peer beyond. Since information cannot travel faster than light, the horizon is a limitation of our knowledge of what is out there.

Second, the standard model of the Big Bang cannot explain why the universe is so close to the critical density. This is referred to as the *flatness problem*. The critical density of the universe is the finely tuned density that would, in some eon, slow the expansion of the universe to a halt but prevent it from collapsing.

Horizon Problem

The cosmic background radiation was detected by accident in 1964 by two scientists, Penzias and Wilson, at the Bell Laboratories in New Jersey. They were constructing a new type of antenna to detect signals from orbiting satellites. The cosmic microwave background noise (or CMB) manifested itself as noise to the annoyance of Penzias and Wilson. Every direction in the sky they pointed the antenna the noise permeated. When you tune your television or radio to an unused channel, the "snow" you see on the screen and the static you hear on the radio is partially attributed to this space noise.

The cosmic background radiation was formed when the energy density of the early universe became low enough for atomic nuclei to capture the free electrons permeating through space. The free electrons absorbed the photonic energy of the early universe making space opaque to photons. In other words, a photon could not travel very far in the young cosmos without getting absorbed by an electron. When the free electrons were captured to form neutral atoms, photons were free to travel unimpeded through space. The universe at this time (about 300,000 years after the Big Bang) finally became transparent. During this epoch, matter and radiation separated or decoupled. This is called photon decoupling.

The same radiation is detected today from all directions in space as the cosmic background radiation, traveling toward us for some ten billion years. This radiation has been stretched (redshifted) to longer wavelengths by the expansion of the universe. This radiation is amazingly uniform. The CMB is the same wavelength looking out in one direction of the cosmos as it is looking out into the opposite direction. The Cosmic Background Explorer (COBE) satellite launched in the early 1990s has finally proved the homogeneity of the background noise. Its smoothness has been determined to be within a few thousandths of a degree. It is often said that space radiates at a temperature of 2.7 degrees above absolute zero (or about -270°C), and on average varies little more than a few thousandths of a degree.

All matter in this early universe was very close together, for the universe at 10^{-34} seconds was an extremely tiny place. Objects of different temperature in close proximity will attempt to reach the same temperature. This is the second law of thermodynamics:

Objects have a general tendency to reach thermal equilibrium.

A hot object (a block of aluminum, for example) sitting on top of a cold object (another block of aluminum) will radiate a portion of its heat energy to the cold object. The cold object will in turn absorb a portion of the hot object's heat energy. This action will cause the hot object to decrease in temperature while the cold object gets warmer. Eventually the two will reach a steady state—the same temperature throughout. This is called "thermal equilibrium."

The cosmic background radiation emanates a blackbody spectrum. This was verified by the COBE satellite. A blackbody is an object that is in thermal equilibrium. The blackbody is said to be "black" because it neither emits nor absorbs heat to its surroundings. Just as the Big Bang model predicts, the cosmic background spectrum is one of a blackbody. This is reasonable (that the universe is in thermal equilibrium) since there are no surroundings (to the universe) to emit or absorb energy.

Entropy is the thermodynamic property that describes the amount of disorder in a system. The universe is slowly reaching a state of increasing disorder. Entropy in the universe is increasing. This is why a cup of coffee sitting on the desk gets cold. The hot coffee (a state of concentrated energy—order) eventually gets cooler (leading to a state of dispersed energy—disorder). This is also why a child's building blocks stacked up on the floor will scatter haphazardly if disturbed. Systems tend to move toward a state of increasing disorder. This is the principle of the increase of entropy. It is highly unlikely that the child's blocks, if smashed into, would rearrange themselves into a model of the Empire State Building!

The universe is believed to be about ten billion years old. Since nothing can travel faster than the speed of light, we cannot possibly see past ten billion light-years. This is said to be the "horizon" of the universe—the barrier that one cannot see past (no matter how powerful the telescope). This barrier corresponds to the era of photon decoupling just before the universe became transparent to photons some 300,000 years after the Big Bang. The universe has a finite age, so there is a maximum distance light could have traveled since the beginning.

Space, ten billion light-years in one direction, is the same temperature as space ten billion light-years in the opposite direction. This seems impossible since, in order to be at the same temperature, the objects must have reached a state of thermal equilibrium long ago. But in order to reach thermal equilibrium, they must have been in causal contact sometime in the distant past. This seems impossible because the objects are too distant to have ever been in contact (assuming that Einstein was correct in stating that nothing can travel faster than light). *This is the horizontal problem.* The universe is larger than the distance light could have traveled since the creation of the universe.

Flatness Problem

The standard model of the Big Bang cannot explain why the universe is so close to the critical density. Density is the relationship of an object's mass compared to its volume. The density of the universe is defined as the total mass of the universe (a huge number) divided by the total volume of space it occupies (also a very large number). A very large number divided by another very large number will equate to a reasonably small number. This is the case here. The average density of the universe is a small number—about 0.2 atoms per cubic meter. This is practically a vacuum.

The universe of today is expanding. Edwin Hubble discovered this in the early part of the twentieth century in the redshift of distant galaxies. Since every particulate of matter is attracted to every other particulate of matter due to gravitational attraction, the universe will eventually begin to collapse under its own gravity unless the average density is less than some critical density. If the average density of the universe is less than the critical density, the universe will continue to expand forever into oblivion. Galaxies would continue to recede from one another, farther and farther apart until they just faded from sight—even from the sight of our most powerful telescopes. The universe would be exiled to a cold, lonely eternity.

On the other hand, if the average density of the universe is greater than the critical density, the universe will eventually halt its expansion and begin to contract into a Big Crunch. The universe after billions of years would shrink smaller and smaller and, as with any contracting closed system, would become hotter and hotter. Galaxies would become closer and closer, eventually filling the sky with a perpetual twilight, continuing through to where the early universe started. Perhaps the Big Crunch would manifest itself as the Big Bang in reverse, cycle back to the beginning of time and produce a stupendous explosion of energy and matter. Here we go again! Perhaps we (as creatures of matter and energy) are riding on one cycle of an endless loop of creation.

If the average density of the universe is *exactly* equal to the critical density, the universe would eventually cease its expansion and halt, never to expand farther nor contract. The universe would become static and exhibit perfect flatness.

The difficulty of a dynamic system (such as the universe) to reach such as state of equilibrium is analogous to the likelihood of a perfectly smooth steel ball rolling up a perfectly smooth glass hill and stopping at the exact summit. This is a highly unlikely event. Cosmologists have defined the density parameter of the universe (Ω–Greek letter omega) as the ratio of the average density of the universe to the critical density. If the density of the universe was equal to the critical density, Ω would equal 1. This seems to be the case for the universe. *This unlikely circumstance is the flatness problem.* The density parameter, Ω, for the universe seems extremely close to 1. If this fact holds true, the universe is flat. It will eventually stop expanding but not collapse. If Ω is truly 1, the universe will halt and remain static. Stars will eventually burn themselves out and either nova or die a silent death, and galaxies (on average) will never come closer nor grow farther apart, and their constituents will also disperse, dying in a seemingly endless sea of radiation.

Modification of the Standard Model

Modification of the standard Big Bang model of the universe, to account for the discrepancies of the horizon problem and the flatness problem, came in 1981 by the American physicist Alan Guth. His theory is called "inflation." It describes the universe from an age of about 10^{-35} seconds to about 10^{-32} seconds. According to Guth the primordial universe underwent extremely rapid expansion by a huge factor for a brief instant, shortly (10^{-35} seconds) after the Big Bang. The universe increased in size by a factor of 10^{50} (100,000,000, 000,000,000,000,000,000,000,000,000,000,000,000). During this extremely brief moment of creation, the particles and antiparticles were formed—all matter/antimatter of the universe was formed.

During the brief period before inflation, the temperature of the universe was able to become uniform, and the universe reached a state of thermal equilibrium. This state of thermal equilibrium was not affected by the rapid inflation. Inflation solved the horizon problem. In the small region before inflation, there was time enough to reach thermal equilibrium. Large-scale uniformity must have been an initial condition. The universe at 10^{-35} seconds must have reached a state of thermal equilibrium before the rapid expansion occurred. Otherwise, the universe would not be as uniform in temperature (detected in the cosmic background radiation), as observed today. Thus, the uniformity of temperature observed today must be a direct result of inflation.

Guth explained that during the inflationary epoch the universe expanded from about 10^{-52} meters (much smaller than a proton) to about 10^{-1} meters (about the size of a baseball) in about 10^{-32} seconds. Since the brief but circumstantial inflationary era, the universe has expanded continuously at its more leisurely pace. Since inflation, the universe has expanded from the size of a baseball to its present size (a factor of 10^{28}) in some ten billion years.

Inflation also solved the flatness problem. Somehow (cosmologists are not quite sure how) the extremely brief, but explosive, burst of inflation forced the density parameter, Ω, to a value extremely close to 1. This anomaly produced the "flat" universe apparent today.

As predicted in Einstein's general theory of relativity, inflation seems to have been caused by some repulsive force acting against gravity. During this extremely brief period, energy was released as space unfurled. Particles and antiparticles were produced forming quarks/antiquarks and leptons/antileptons. This process (known as "baryogenesis") produced all of the matter/antimatter in the universe today—even the quarks and electrons that constitute your body, your clothes and your dog!

EVIDENCE OF BIG BANG THEORY

It is almost inconceivable to believe that all that is and was and is to come began from an infinitesimal point smaller than a proton—smaller than the smallest particle—from a cataclysmic release of energy long ago. Studying cosmology—the science of the extremely large and the extremely small—to learn more about our origin and of the origin of the universe around us, is an ancient human tradition probably older than recorded history itself. Many ancient cultures have built their civilizations around an astronomical calendar. Numerous ancient cities were constructed to conform to significant astronomical events or alignments such as the summer solstice or the vernal equinox. The ancient Egyptians were fantastic mathematicians and astronomers. The Pyramids of Giza constructed over four thousand years ago align perfectly with the three stars of Orion's belt. One pyramid is even offset slightly with the other two just as the western-most star is slightly askew with the other two in the sky.

Our inquisitive nature is part of who we are as curious creatures. We will someday inch closer and closer to a clearer understanding of space and time itself—the "how" of genesis. We may never in our mortal existence understand the "why." That question (or at least the answer) is reserved for religion and philosophy. We are not permitted, as mortal creatures, to know the mind of God. Having a clearer understanding of our origin and the formation of the universe does not make us any less spiritual, it only tells us that we must look deeper—further than the creation of space and time to deep within ourselves.

If the universe is expanding, it follows that everything was closer together earlier in the past and that the universe definitely had a beginning. This is the basis of the Big Bang theory.

The Big Bang is the theory that the universe began as a stupendous explosion (words hardly do this event justice!). From a primordial point smaller than an atom, smaller than a proton or anything imaginable, all galaxies, planets and people came to be. The Big Bang occurred approximately ten billion years ago. At that instant, spacetime unfurled. Three spatial dimensions (length, width and height) unfolded, as did a fourth dimension—time. In essence the universe was created from nothing. The net energy of the universe is zero. Matter may be considered positive energy. After all, matter (protons, neutrons, electrons, and so on) is simply a manifestation of tiny trapped packets of energy. All particles of matter are attracted to all other particles of matter by gravity. Gravitational potential between two particles of matter is the embodiment of negative energy. So all of the positive energy (manifested as matter) and all of the negative energy (manifested as the gravitational potential—gravity) summed together results in a net total energy of the universe of zero. The law of conservation of energy is preserved.

The universe has been expanding ever since the Big Bang, and the tremendously hot, bright blast has dwindled down to a dim, cosmic whimper detected as the cosmic background radiation—the echo of a violent birth. The Big Bang theory of the universe is a beautiful model that describes the genesis of the universe. Several pillars support its platform:

- the expansion of the universe

- the cosmic background radiation

- the abundance and even distribution of helium throughout the universe

Expansion of the Universe

In 1929 Edwin Hubble deduced, through careful observation of distant galaxies, that the universe is not static. The universe is indeed expanding. The fabric of the universe is spreading in every direction, simultaneously in all four dimensions—three of space and one of time—the spacetime continuum. Hubble found that distant galaxies are receding from our galaxy, on average, with a speed equal to the product of the Hubble constant and the distance to the galaxy. This is known as Hubble's law and marks the cornerstone of modern cosmology. The Hubble constant describes the current expansion rate (recession velocity per unit distance) of the universe. It is independent of distance, but decreases with time, as universal expansion is slowed by gravitational attraction. The Hubble constant is not truly a constant. It is slowly decreasing with time, as gravity consistently pulls back.

Distant galaxies are moving away from us at a speed proportional to their distance. Hubble found that a galaxy's redshift is directly related to its distance, and he found that the more distant the galaxy, the greater its redshift. This simply means that every galaxy seems to be moving away from every other galaxy, and the farther away a galaxy is, the faster it seems to have been traveling. Space is expanding.

It is a reasonable assumption that if all galaxy clusters are moving away from all others, then in the past they must have been much closer together. In the distant past (ten billion years or so) all matter must have been contained in a much smaller space. The expansion of space must have been the result of a primordial release of energy. Everything came into being from a point pregnant with potential in which the dimensions of space and time unfurled. The expansion of the universe is a direct result of the Big Bang. It is a major supporting pillar that justifies the Big Bang model of the universe.

Cosmic Background Radiation

If the universe started from a stupendous explosion, might it be possible that we may still be able to "hear" the echo of the distant rumble? Might we still feel the warmth of the fires of creation? We can—sort of. Intergalactic space—the vast lonely voids between galaxies—is not completely cold. The frigid warmth of space (about 2.7 degrees above absolute zero) is saturated by the smoldering embers of creation. Absolute zero is the absolute lowest temperature possible in nature—the point at which all motion, even molecular vibration, stops. Space is permeated with microwave radiation—photonic remnants of a cataclysmic explosion some ten billion years ago. These photons in the microwave region of the spectrum are the same radiation that cooks food in microwave ovens, only much less intense.

The cosmic background radiation (discovered by Arno Penzias and Robert Wilson in 1964) has been losing energy since its debut just after the Big Bang some ten billion years ago. As the universe expands so does the wavelength of the cosmic microwave background (CMB). The wavelength of the background radiation has been stretched farther and farther, and is now in the millimeter range, which corresponds to the microwave region of the spectrum.

The cosmic background radiation was not released at the instant of the Big Bang but, rather, hundreds of millennia later. The universe, shortly after creation, existed as an extremely hot, energetic soup of matter and energy. Radiation was scattered and absorbed by the free electrons that permeated the soup. Radiation (photons) emitted shortly after the Big Bang traveled a very short distance before being reabsorbed by the free electrons. When the universe became cool enough to allow the free electrons to be captured by the plasma of atomic nuclei, the photons released were able to travel through space unimpeded. Here radiation separated or decoupled from matter. This is known as the era of photon decoupling, and occurred some 300,000 years after the Big Bang. The universe finally became transparent to photons, and the photons (now "visible" as the cosmic microwave background) have been traveling through space and time ever since. Remember that the universe is a tremendously vast expanse, and light has a finite and constant speed.

Detecting this radiation is to observe the universe as it existed approximately 300,000 years after the Big Bang. This is what the Cosmic Background Explorer (COBE) satellite did in the early 1990s. COBE measured the wavelength of the cosmic background radiation across the sky. It is amazingly homogeneous from every direction peering out into the cosmos. This fact leads to reason that all matter in the universe came to a state of thermal equilibrium long ago. In order to reach thermal equilibrium all matter must have been in causal contact long ago.

If all matter was in contact in the distant past, it must have started expanding from a very small point.

COBE showed that the cosmic background radiation exhibits a blackbody spectrum. A blackbody produces a distinctive curve when the intensity of the radiation emanating from the object is plotted versus its frequency. A blackbody radiation curve (energy density versus frequency) is produced when an object has established a state of thermal equilibrium with its environment just as a furnace that has been constantly burning for a long time does. The furnace that is the universe has burned steadily for some ten billion years, so it is safe to say that the radiation left over from the Big Bang has reached thermal equilibrium. If the cosmic background radiation is indeed the remnants of a cataclysmic fireball, it should exhibit the properties of a blackbody. It does.

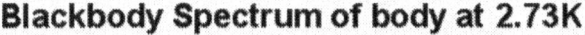

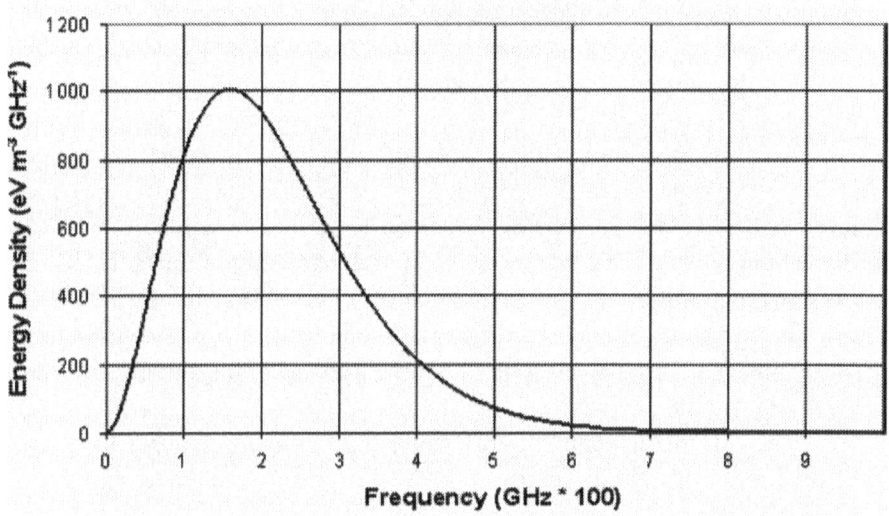

A blackbody is an object which absorbs any and all radiation incident upon it, and releases electromagnetic energy across the spectrum. A blackbody has established thermal equilibrium with its surroundings. It emits radiation at wavelengths and intensity depending on its temperature. The characteristic curve of the intensity of the emitted radiation versus frequency is called a blackbody spectrum. The Cosmic Background Explorer (COBE) satellite measured such a characteristic curve in the cosmic background radiation. The energy distribution detected by the COBE satellite agrees with quantum theory for blackbody radiation. The measured curve follows the Planck radiation formula. This means that the universe established thermal equilibrium long ago which further implies that all matter in the universe started from a tightly confined space.

Any object above absolute zero emits photons (electromagnetic radiation). Bodies at different temperatures emit radiation of different wavelengths. A blackbody gives off energy at a rate proportional to its temperature. There is no such thing as an object that emits no radiation. Even common, everyday objects, such as the sofa or chair you may be sitting on or the wall behind you, radiates energy (primarily in the infrared region of the electromagnetic spectrum invisible to our eyes). Objects that are hot emit a great deal of infrared radiation, and if energetic enough, the object will begin to emit visible light—first glowing deep red, then orange, and perhaps white hot. This light, if decomposed into its constituent wavelengths by passing it through a prism, will produce characteristic peaks in intensity at certain wavelengths. Different materials produce a characteristic spectrum of lines when its light is decomposed. Using this phenomenon, astronomers are able to distinguish certain elements within distant starlight. The composition of a star or distant galaxy millions of light-years away may be revealed using this technique, discerning individual and distinct elements within the light.

An object that has reached thermal equilibrium will absorb electromagnetic radiation in the same wavelengths that it is emitting. Hence, a steady-state condition is established. The spectrum of such an object in thermal equilibrium will no longer exhibit characteristic lines. The graph of the intensity of the radiation versus its frequency will instead produce the smooth curve characteristic of a blackbody. The spectrum produced by the blackbody radiation is independent of whatever material the glowing object is composed of. The thermal spectrum depends only on temperature.

All objects in thermal equilibrium exhibit this blackbody radiation and produce the familiar blackbody spectrum. As COBE proved, the universe is no different. The universe is a tremendous blackbody that established a state of thermal equilibrium long ago, which supports the notion that all matter must have been in causal contact at its birth. The cosmic background radiation is the radiation emitted by this stupendous blackbody universe. The existence of the cosmic microwave background and the fact that it produces a characteristic blackbody spectrum is another major proof of the Big Bang theory of the universe.

Universal Abundance of Helium

Moments after the Big Bang, the universe burned hotter than the center of the sun. The explosive expansion prevented atomic nuclei from fusing higher up the periodic table past deuterium and helium (besides a minute trace of lithium). In those first three minutes of creation, approximately 23 percent of the matter in the universe (in the form of hydrogen) transformed into helium. That percentage of helium is distributed somewhat evenly throughout the universe.

Young stars polluted with stellar material, expelled from ancient supernovae explosions, contain considerably more of the heavier elements than do very old stars. Stars are born, process their hydrogen fuel into helium, carbon, oxygen, nitrogen, silicon and on up to iron, and eventually (if massive enough) conclude their lives in a stupendous explosion called a supernova. Supernovae spread the heavy elements synthesized in the cores of massive stars just as an earthly wind spreads airborne seeds and spores so that life may continue somewhere far away.

The heavy elements, thrown free of the blast, eventually find their way to some distant cloud of interstellar hydrogen gas and accumulate gradually. The cloud eventually begins to condense, and eons later a young star is born. For this reason, young stars are rich in heavy elements. The elements heavier than hydrogen and helium have already been processed by some other ancient stellar furnace and have made their way to some distant interstellar cloud of hydrogen gas where the stellar cycle repeats.

The intense gravitational pressures at work deep within a star prevent the outer layers from mixing with the core material. Therefore, the material contained in a star's outer layers, which was not processed by the core, was present in the interstellar gas cloud from which the star was formed.

It may seem logical to assume that very ancient stars (billions of years older than the sun—formed when the galaxy was young) contain only a very small percentage of helium. Surprisingly, this is not the case. Even very old stars seem to be composed of almost one-quarter helium. This much helium could not have been produced by the star's own nuclear reactions. Apparently our young galaxy did not begin its life as hydrogen only. The evidence supports the theory that the galaxy started from an inventory of hydrogen and about 23 percent helium. The existing stars could not have produced the abundance of helium observed today. All of the stars in the universe could have produced only a fraction of all of the helium present at the rate they currently produce energy by fusing hydrogen into helium.

Light from distant stars (and galaxies alike), analyzed with mass spectrometers, shows that approximately 23 percent of visible matter in the universe is composed of helium. This fact supports the theory that the helium was formed in the beginning from the cataclysmic explosion that started it all. This would also explain why helium is so abundant and so evenly distributed throughout space. If it all started at once, in a contained volume much smaller that its present greatness, it is a logical assumption that the percentage of helium would be homogeneous throughout the entire volume. Therefore the percentage of helium in the matter that constitutes the universe, and its even distribution, is a major pillar that supports the Big Bang model of the universe.

The abundance of deuterium (an isotope of hydrogen containing one neutron) is somewhat of a mystery. Deuterium is a by-product of nuclear fusion, and is destroyed in stars rather than created. Deuterium is easier to burn in stars (via fusion) than its little sister hydrogen. So, any deuterium present in young stars would be consumed first, before the star settled down to its lengthy hydrogen-burning stage. Yet, there is a small percentage of deuterium proliferating throughout the universe. The highly probable explanation for the existence of deuterium throughout the universe is that is was leftover as a by-product of the helium fusion process in the nucleosynthesis era during the first three minutes after the Big Bang. The existence of deuterium is another piece of evidence supporting the Big Bang model of the universe.

Major evidence exists supporting the theory that the universe had a beginning of extremes—the expansion of the universe discovered in the redshift of distant galaxies, the existence of the cosmic background radiation and the fact that it exhibits a blackbody spectrum, and the abundance of helium and deuterium and their even distribution throughout the universe. All of these facts together mold an elegant model of the Big Bang.

The Big Bang theory of the universe is just that—a theory. It is not a law—yet. There may be other explanations for these anomalies. Some cosmologists dispute the theory, while a whopping majority support it. How else would one describe the distant past if all matter and space is racing away from us and all other? It seems obvious, but as scientists, and sometimes objective creatures, we must not rule out other possibilities. Science and religion must stop disputing. Whenever one gains a bit of insight on the surrounding universe (as Galileo and Copernicus once did), it does not make him or her heretic. Rather, it paints a beautiful picture of existence—one that praises Supremacy instead of simply dismissing it

EXPANSION AND MISSING MATTER

The universe is expanding. As time passes, the universe grows larger and larger. Will it expand forever without bound, becoming darker and colder? Will it eventually halt? Will the universe halt and begin to collapse with galactic clusters moving closer and closer, eventually leading to a Big Crunch? Astronomers and cosmologists are unsure. The central theme of cosmic expansion (or collapse) is gravity and mass density.

The density parameter of the universe (Ω) is the ratio of the average density of the universe to the critical density. The critical density is that density which is sufficient enough to cause the universe to slowly halt expansion as a result of gravitational attraction, but insufficient to initiate collapse. The current tally of *luminous* matter in the universe puts Ω at 0.04. This implies there is insufficient matter to prevent the expansion of the universe from continuing forever into a cold, dark and lonely eternity. This number (0.04) seems far from unity (only 4 percent) when dealing with numbers and scales from everyday life, but when considering the unbelievable vastness and grandeur of the universe and the amount of matter in the cosmos, this parameter, Ω, is amazingly close to one. The density parameter could have easily been 10^{-12} or 10^8, but luckily for us, and for other citizens of the cosmos, it is not.

The fact that the universe is so close to the critical density exemplifies an extremely unlikely condition. The necessary precision is uncanny, almost as if this parameter, Ω, was tuned by design. This unlikely circumstance in the standard model of the Big Bang is called the flatness problem.

The standard model of the Big Bang cannot explain why the universe is so close to the critical density. The universe is on the verge of eternal expansion and a crushing death. Stated another way, the universe will either endure an eternally dark, cold and lonely existence or die an intensely hot, crushing death. The density of the universe will decide its fate.

If Ω was too small (less dense) early in the universe's infancy, then gravity would have been insufficient to conglomerate miniscule densities in the early universe to form stars and planets, galaxies and superclusters of galaxies. Life would have never been able to form. The universe would have become a vast, dark and uninteresting cloud of hydrogen and helium gas. If Ω was too large (meaning the early universe was sufficiently dense), the universe would have collapsed back on itself long ago, and we would never have been here to contemplate it. In this scenario, life would have been unable to form, as well. There would have been insufficient time. One can appreciate the importance of the values of the density parameter (Ω) and the gravitational constant of the universe (G). The necessary precision required to produce the relatively stable cosmos evident today points to two possibilities: either the universe in which we are tenants happens to be one of countless trials of failed universes, until the numbers finally coincided, or the numbers came to be by design.

The fate of the universe lies with the amount of mass it contains. Cosmologists have an educated grasp on the size of the universe, but in order to determine its density, mass must be determined. How much matter is contained in the universe?

Determining the amount of matter a distant galaxy contains is a difficult undertaking. Stars illuminate with different and sometimes variable intensities. Not all matter is luminous—dust clouds, for example, are dark. One indication of a galaxy's mass is revealed in its rotation. The angular velocity (spin rate) of a gravitationally bound disk is dependent upon its mass. If the disk rotates too rapidly, it will fly apart like a high-speed mixer slinging cake batter! If the disk rotates too slowly it will collapse in on itself due to gravitational attraction.

The angular velocity of a rigid disk, such as a flywheel, a Ferris wheel or a Frisbee, does not depend on its mass since it is held together by it internal structure. A galaxy, however, is not rigid. There are no firm supports or beams that bind one star to the next. Stars within a galaxy are bound by a force—gravity. Gravity pulls all stars toward the center of their parent galaxy—the center of mass—and the centrifugal force induced by galactic rotation pulls stars away from the center of rotation. In order for a galaxy to maintain its structure, the force of gravity pulling all of the stars toward the center must be equivalent to the centripetal force required for the stars to hold orbit. These opposing forces remain in balance and maintain the structural integrity of the gravitationally bound disk (galaxy). The angular velocity of a galaxy (viewed edge-on) can be easily calculated from the difference in Doppler shift on each side. The receding side of the galaxy will be slightly redshifted and the advancing side will be slightly blueshifted. By comparing the differing velocities of the two sides, a reasonable estimate of the galaxy's rotation rate can be calculated.

Using physics and a bit of intuition one would expect a galaxy to rotate with similar dynamics as a planet orbits its parent star. A planet with a smaller orbit must revolve around its parent star faster in order to maintain orbit. Likewise, the same planet with a larger orbit must revolve slower around its parent star in order to remain in orbit. The fact that an orbiting planet must revolve faster the closer it orbits its star and slower if farther out is known as Kepler's Third Law.

Kepler's Third Law describes the relationship between the orbital velocity of a body (such as a planet) and its distance to the system's center of mass (such as a star). The law holds true for systems whose mass is concentrated within the orbit, closer to the center. One would expect the constituent stars of the spiral arms of a galaxy to revolve with similar predictability—a star farther from the galactic center should revolve slower than a similar star closer to the center. This rule does not seem to apply to a majority of galaxies. Stars in the outer perimeters of the spiral galaxies seem to revolve around their galactic center much too fast to maintain the structural integrity of the galaxy. Cosmologist Vera Rubin discovered this anomaly. A vast majority of galaxies should fly apart, but they do not. The internal motions of galaxies suggest the presence of some form of unseen matter holding them together. The anomalous internal motion of galaxies signifies the presence of a huge portion of invisible matter outside of the spiral arms, perhaps in some sort of galactic halo.

If the solar system was surrounded by a halo of matter—enough matter to equal the mass of the sun—then the outer planets, such as Jupiter and Neptune, would orbit the sun just as fast as the inner planets, such as Earth. Astronomers and cosmologists have termed this unseen undetected substance "dark matter." Evident from their motion, galaxies must be encased in some form of dark matter. The phenomenon is characteristic in galactic rotation. Stars in the outer portions of spiral arms seem to revolve around the galactic bulge just as fast as the inner stars.

What form does this mysterious matter take? Either a majority of the universe is made of dark matter or we have made a gross mistake applying Newton's, Kepler's and Einstein's equations for force and gravity. There are theories that suggest the form of the missing mass—the very large and the very small.

One theory describes a massive halo surrounding our galaxy and similar ones around other galaxies. The constituents of these massive clouds are likely to be small, unseen dwarf stars probably formed early on in the evolution of galaxies. These MACHOs (or MAssive Compact Halo Objects) may exist in the form of brown dwarfs—objects with too little mass to shine—and white dwarfs—dense burned-out cores of dead stars. MACHOs are described as massive not because of their size, but because they contain mass and they would have to be very numerous to affect galactic rotation.

Evident from their rotation, galaxies could be surrounded by clouds of dark, cold objects just as the Oort Cloud of icy comets surrounds our solar system. If this cloud of MACHOs surrounds galaxies like some form of halo, and if a large portion of the mass exists throughout and well beyond the spiral arms, this would explain the phenomenon of excessive galactic rotation. Gravity induced by the excess mass well beyond the central galactic bulge would lessen the centrifugal force opposite the galactic disk and allow it to rotate faster without flying apart.

Another theory that may explain the curious motion of galaxies pertains to the very small, the subatomic. This theory suggests that WIMPs (or Weakly Interacting Massive Particles) that bathe the universe in the wake of the Big Bang may contain the slightest quantity of mass. The average mass density of the universe, if all atoms were evenly distributed in a giant gas cloud throughout its entirety, is about 0.2 to 0.5 atoms per cubic meter. This equates to about one particle per volume the size of a small closet. The average particulate matter density of the universe is extremely low, but the density of weakly interacting massive particles (neutrinos) is comparatively higher.

Just as the Big Bang produced the microwave background radiation emanating from all points in space, the creation of space and time also produced a similar background of neutrinos. The background radiation emanates with over a billion photons for every proton and neutron contained in the universe. Similarly, the neutrino background emanates with almost a half billion neutrinos for every proton and neutron. In other words, for every atom in the universe there are hundreds of millions of neutrinos.

Photons are massless. They contain no mass and therefore contribute nothing to the mass density of the universe. Up until recently neutrinos were also thought to contain no mass. However, if neutrinos exhibit even the slightest amount of mass, this may account for the missing mass of the universe, and the excess mass inventory may be sufficient to close the universe.

Neutrinos are released during nuclear reactions, like those that occurred when the universe was approximately one second old, and those still occurring in the cores of stars. Considered "weakly interacting," neutrinos respond only to the nuclear weak force (responsible for nuclear decay). This fact makes them extremely difficult to detect and even harder to quantify. At this very moment, hundreds of millions of neutrinos from the sun are streaming unimpeded through your body and even through the earth. In fact it would take several light-years of solid lead to stop even one of them.

To search for dark matter is to seek the missing mass of the universe and to explain the curious rotation of galaxies. The amount of visible matter in galaxies is insufficient to explain their rapid rotation. They should fly apart. MACHOs and WIMPs may be responsible for these anomalies. Depending on the content of matter, the universe will expand forever or cease and fall in reverse. The universe will fade into a dark oblivion, dying a slow, cold and lonely death or, if there is sufficient mass, the universe will die a comparatively shorter, hot, crushing death.

AFTERWORD

Cosmology is a perplexing subject. The evolution of the universe remains unexplored and ignored by the general audience, but everyone has caught himself or herself gazing up at the night sky pondering the beauty and vastness of space. Cosmology is a difficult subject, but becomes a little clearer once certain key points are understood:

- The universe is a tremendously vast, dark expanse of space and time—referred to as the spacetime fabric of the universe.

- The universe had a beginning; it has not existed forever.

- The universe was initiated by a tumultuous, violent event called the Big Bang.

- The universe began from an infinitesimal point—a point smaller than a proton—and from this point, space and time unfurled. Everything originated from this point.

- The universe is estimated to be somewhere between 10 to 15 billion years old.

- The universe is expanding.

- The structure of the large-scale universe is summarized by the cosmological principle—on the large scale, the universe is evenly distributed and appears the same from any direction. Stated concisely, the universe is homogeneous and isotropic.

- Evidence supporting the Big Bang exists today:

  - The expansion of the universe.

  - The existence of the cosmic background radiation.

  - The abundance and even distribution of helium throughout the universe.

- The large-scale universe is organized by constituent building blocks—galaxies—just as matter is arranged by its constituent—atoms.

- The elements (like carbon, oxygen, nitrogen and iron) that constitute matter—from puppy dogs to the rings of Saturn—were synthesized in the center of stars—within stellar cores.

- There are several numbers important to cosmology:

  - Ω—Density parameter of the universe—ratio of the actual density versus the critical density of the universe. Omega (Ω) is important to cosmology because it gives clues as to the ultimate fate of the universe—eternal expansion, eventual halt or a Big Crunch.

  - λ—Cosmological constant—term representing a repulsive force preventing the universe from collapsing under its own gravitational influence. Lambda (λ) is important to cosmology because it also gives clues as to the ultimate fate of the universe by affecting the expansion rate of the universe.

  - Q—Factor in determining the structure of the universe—determines the homogeneity of the universe evident in the cosmic background radiation. Q is actually the ratio of two energies. Q is important to cosmology because it gives clues as to the texture of the early universe, and gives evidence to explain the structure of the large-scale universe.

  - H_0—Hubble Constant—a description of the expansion rate of the universe. H_0 is important because it gives clues as to the age of the universe.

- There is room for religion in science, and there is room for science in religion. The two are not mutually exclusive; there should be no controversy. The authors of the Bible did not intend it to be a textbook of science, so there should be no conflict between faith and modern science. Whenever the scientific community gains a bit of insight on the surrounding universe, it does not squeeze out religion, rather, it paints a beautiful picture of existence, one that praises Supremacy.

GLOSSARY

absolute zero—Zero Kelvin (-273 °C). The lowest achievable temperature.

Alpha Centauri—The closest star system to the solar system. The Alpha Centauri system is 4.3 light-years distant from the earth.

Andromeda Galaxy—The sister galaxy to the Milky Way, similar in size, shape and number of stars. The Andromeda Galaxy is approximately 2.2 million light-years distant and is the nearest spiral galaxy.

anthropic principle—Notion that limits fundamental constants of the universe, and constrains cosmological theories to fit conditions necessary for the existence of life.

antimatter—The "mirror image" of matter where particles have the same mass and spin but opposite charge. The union of matter and antimatter would result in complete and total destruction of each constituent—transforming the matter/antimatter into energy. The antimatter equivalent of an electron is a positron, and the antimatter equivalent of a proton is an antiproton. For every matter particle there is an equivalent antiparticle.

atom—The smallest building block of matter.

Big Bang—The theory that the universe originated from an infinitely small point somewhere between ten and fifteen billion years ago. From this event, space and time were created, along with all matter and energy in existence. The term "Big Bang" was coined by the English physicist Sir Fred Hoyle.

Big Crunch—The ultimate fate of a closed universe, where the average density of the universe is greater than the critical density ($\Omega > 1$). In the Big Crunch scenario, the universe contracts further and further eventually reaching a Big Bang singularity. The Big Crunch is similar to a Big Bang in reverse.

black hole—The end result of the collapse of an extremely massive star. The gravitational field is so strong that nothing, not even light, can escape.

blackbody—An object that has reached a state of thermal equilibrium thus emitting a blackbody spectrum.

blackbody spectrum—Characteristic radiation curve plotted as energy density versus frequency emitted by a body that has reached thermal equilibrium.

blueshift—A phenomenon evident in stars and galaxies in which their spectra are shifted toward the blue end of the electromagnetic spectrum. The effect is due to the object's relative motion toward the observer's perspective. See *redshift*.

boson, weak gauge—Particle carrying the weak nuclear force.

bosons—Family of particles that carries the four fundamental forces. These "force particles" include the gluon, photon, weak gauge boson and graviton.

Brownian motion—The random motion of particles in a fluid directly affected by temperature.

Chandrasekhar limit—Mass equivalent to 1.44 solar masses. Stars more massive than the Chandrasekhar limit will eventually turn into a supernovae and transform into neutron stars after they exhaust their supply of hydrogen fuel.

closed universe ($\Omega > 1$)—A universe whose density is greater than the critical density so that it will eventually collapse in on itself.

cluster, galactic—Dense conglomerations of galaxies stretching some tens of million of light-years and constituting hundreds to thousands of galaxies. Clusters are the largest gravitationally bound structures in the universe, meaning that they are not stretched apart by the expansion of the universe.

comet—A small body (smaller than a planet) containing mostly rock, metal and ice in orbit (usually highly elliptical) around the sun. When comets come in close proximity to the sun, the ice vaporizes leaving a long trail of water vapor always pointing away from the sun due to the action of solar wind particles. This long trail of water vapor can be seen from the earth, as the comet's tail. Comets are the remnants of the formation of the solar system.

cosmic microwave background (CMB) or cosmic background radiation—The remnants or afterglow of the Big Bang resulting from the release of photons. This radiation permeates the entire universe and, due to redshift caused by the expansion of the universe, its energy level has been reduced to an average temperature of 2.73 Kelvin.

cosmological constant (λ)—The term introduced into the general theory of relativity by Einstein to prevent the universe from collapsing under its own gravitational influence. Einstein believed that the universe was static. The term is symbolized by lambda (λ) and represents a repulsive force. Modern theory agrees with the introduction of λ, but Einstein introduced it for the wrong reason and later shamefully retracted it.

cosmological principle—The notion that the universe is homogeneous (evenly distributed) and isotropic (looks the same in all directions).

cosmology—The study of the universe on a grand scale. Cosmology encompasses detail within the universe from the extremely minute (subatomic particles) to the stupendously gigantic (supercluster complexes of galaxies), and combines the fields of astronomy, astrophysics and particle physics.

cosmos—The universe as a whole and everything contained within it.

critical density (of the universe)—The density (ratio of mass per unit volume) sufficient to eventually halt the expansion of the universe, but insufficient to cause it to collapse in on itself.

dark matter—An unseen and so far undetected form of mater theorized to account for over 90 percent of the matter in the universe. Dark matter is said to be "dark" because it is not visible. It does not radiate electromagnetic energy, or at least dark matter does not emit enough energy to be detected by modern means.

deuterium—An isotope of hydrogen containing one neutron.

Doppler shift (or Doppler effect)—The effect caused by an object's motion relative to an observer in which waves (either electromagnetic or audible) emanating from the object are stretched if the object is receding, or compressed if the object is approaching the observer.

electromagnetic force—One of the four fundamental forces of nature, it is responsible for attracting particles with opposite charges and repelling particles with like charges. The electromagnetic force is responsible for producing electromagnetic waves in which visible light is composed. The electromagnetic force becomes weaker with distance, but is infinite in range.

electromagnetic spectrum—The entire collection of electromagnetic waves.

electromagnetic waves——Energy carried by photons—the combination of electric and magnetic fields. Electromagnetic waves range from radio waves, microwaves, infrared, visible light, ultraviolet, X-rays, gamma rays.

electron—A negatively charged particle in orbit around an atomic nucleus.

energy—The ability to do work—force realized over a distance. Also, according to special relativity, energy is another manifestation of mass: $E = mc^2$.

entropy—Thermodynamic principle that describes the amount of disorder in a system. Entropy in the universe is increasing which, stated another way, means the universe is slowly reaching a state of increasing disorder.

escape velocity—The speed an object must attain in order to leave the gravitational influence of another body (like a planet, moon or star). The escape velocity for the earth is approximately 25,000 mph.

Euclidean geometry—The geometry of flat-space where the sum of the angles of a triangle equals exactly 180°, and two infinite parallel lines never diverge nor recede.

event horizon—Area around a black hole defined by the Schwartzchild radius.

ex nihilo—Literally, out of nothing. In reference to cosmology the term describes the creation of the universe from no energy or mass input—from nothing.

false vacuum—A quantum state where the sum of matter and energy equates to zero. From this state, particles can seemingly appear from nowhere when they are created from the energy flux.

fermions—Family of particles that constitute matter. Quarks and leptons are members of this classification—the "matter particles."

fission—Nuclear reaction that splits the nuclei of atoms apart. The process of nuclear fission fuels nuclear reactors in power plants and is the destructive force fueling a nuclear weapon.

flat universe ($\Omega = 1$)—A static universe (neither expanding nor contracting).

fundamental forces—(From weakest to strongest) gravity, weak nuclear force, electromagnetic force, strong nuclear force.

fusion—Nuclear reaction which joins the nuclei of atoms together. The process of nuclear fusion fuels the sun.

galaxy—A large, self-contained, gravitationally bound collection of stars. Galaxies, many containing billions of stars, are the building blocks of the large-scale universe—the "atoms" of universal structure.

geocentric—The belief that the earth is at the center of the universe. In this view the sun, planets and stars all orbit the earth. This egotistical theory was accepted by the Christian Church for more than a millennium and any other idea was deemed heresy.

gluon—Particle carrying the strong nuclear force.

grand unified theory—Theory that unites the strong force, electromagnetic force and the weak force. This condition of force unification existed in the first fraction of a second after the Big Bang.

graviton—Particle carrying the gravitational force.

gravity—One of the four fundamental forces of the nature, it is responsible for attracting every body possessing mass to every other body possessing mass. Gravity's range is limitless.

heliocentric—The idea that the sun is at the center of the universe. In this view the earth and planets orbit the sun. This theory was proposed by Copernicus and supported by Galileo.

helium—Element with atomic number 2, containing two protons and two neutrons in its nucleus.

Hertzsprung-Russell diagram—Graph depicting the absolute magnitude of a star versus its spectral class.

homogeneous—Evenly distributed. In this context, on the large-scale, the density of the universe appears relatively constant. See *cosmological principle*.

Hubble constant (H_0)—A measure of the current expansion rate of the universe. H_0 is a description of velocity per unit distance. The Hubble constant is measured in $km \cdot s^{-1} \cdot Mpc^{-1}$

Hubble law—The fact that the farther away a galaxy is, the greater its redshift will be. In equation form, $v = H_0 d$, where v is the velocity of the distant galaxy, d is the distance to the galaxy and H_0 is the Hubble constant.

hydrogen—The simplest element. Hydrogen contains one proton.

isotropic—Looks the same in all directions. In this context, on the large-scale the universe appears the same in all directions to an observer at any location. See *cosmological principle*.

kinematics—The study of motion inspired by Newton's laws.

leptons—Classification of particles that includes electrons, neutrinos and muons.

light-year—A unit of distance equivalent to approximately six trillion miles. The distance light travels in a vacuum in the time span of one year.

MACHOs (or MAssive Compact Halo Object)—Theoretical form of dark matter in the form of countless tiny, unseen dwarf stars congregated into a gigantic halo around each galaxy. MACHOs are described as "Massive" because each object contains mass and "Compact" because they are relatively small compared with ordinary stars.

main sequence star—A star that resides along a long, thin band from the lower-right corner to the upper-left corner of the Hertzsprung-Russell diagram. This band depicts the life cycle of a majority of stars in the galaxy. The sun is a main sequence star.

mass—The amount of substance contained in an object.

matter—Any substance that occupies space and has mass.

Milky Way Galaxy—The galaxy in which our solar system is contained.

neutrino—Neutral particle that has no mass (or seemingly no mass) that travels unimpeded through matter and space. Neutrinos may travel through a light-year of solid lead like photons through the vacuum of space.

neutron—A neutral particle within the nucleus of an atom.

neutron star—A star formed when the electrons and protons composing the stellar core are smashed together, forming neutrons due to the overwhelming density. Neutron stars are extremely dense—a spoonful of matter would weigh more than a mountain here on earth. Neutron stars are small with a radius of about ten miles. The Pauli exclusion principle prevents further collapse by preventing the neutrons from moving any closer. In a sense, neutron stars exist as a gigantic, ten-mile wide, atomic nucleus composed almost exclusively of neutrons and a very thin iron crust.

Olbers' Paradox—The apparent incongruity that the night sky is dark even though there are an immense number of ancient stars and galaxies shining through the universe. Heinrich Olbers contemplated this mystery in the early nineteenth century.

omega (Ω)—The density parameter of the universe. The ratio of the actual density of the universe to the critical density. $\Omega < 1$ describes an open universe or one that will expand forever. $\Omega > 1$ describes a closed universe or one that will eventually collapse in on itself. $\Omega = 1$ describes a flat, static universe.

Oort cloud—Conglomeration of billions of comets encompassing the solar system.

open universe ($\Omega < 1$)—A universe that will expand forever.

parsec—A unit of distance equivalent to approximately 3.26 light-years. A parsec (or parallel second) is the distance to a star that forms a triangle between the sun and the earth forming an angle equal to one second of arc.

Pauli exclusion principle—The dictum that describes how no two particles can simultaneously occupy the same quantum state.

photon—Particle carrying the electromagnetic force. Photons contain no mass.

photon decoupling—The separation of photons from the matter particles some 300,000 years after the Big Bang. Before then, photons that were emitted traveled only a short distance before being absorbed by another matter particle, due to the high density of the universe. For the first time, the universe became transparent to photons. The era of photon decoupling produced what is now observed as the cosmic background radiation.

proton—A positively charged particle within the nucleus of an atom.

pulsar—A rapidly spinning neutron star that beams radio waves into space.

quantum mechanics—The study of subatomic systems.

quarks—Particles that constitute protons and neutrons. Protons consist of two up-quarks and one down-quark. Neutrons consist of two down-quarks and one up-quark. Quarks exist in triplets under normal circumstances.

quasar—"Quasi-stellar" object. A quasar is an extremely energetic object billions of light-years distant; thought to be the early formations of galaxies.

red giant—A star that has exhausted most of its hydrogen fuel in its core and utilizes helium fusion. The helium fusion requires more heat and pressure in the core than the previous stage of hydrogen fusion thus pushing out the outer layers of the star into the surrounding space. The star swells tremendously. The star's outer layer becomes cooler due to the expansion, therefore becomes redder in color. In about five billion years the sun will have exhausted its hydrogen supply and begin its red-giant phase, and its surface will reach to about the orbit of Venus.

redshift—A phenomenon evident in stars and galaxies in which their spectra are shifted toward the red end of the electromagnetic spectrum. This is due to the star or galaxy receding away from the observer's perspective, which stretches the wavelength of radiation in the direction of the observer. The phenomenon is due to the Doppler shift. Almost all distant galaxies are redshifted. In 1929 Edwin Hubble correlated this redshift of distant galaxies to the expansion of the universe.

relativity, general theory of—Einstein's theory proposed in 1915 that describes the spacetime fabric of the universe as being curved or warped due to massive bodies (like stars, planets and galaxies). The general theory of relativity is the modern theory of gravity.

relativity, special theory of—Einstein's theory proposed in 1905 that describes moving references frames. This theory proposed that the speed of light is constant; that energy and mass are different manifestations of the same entity.

Schwartzchild radius—Radius of an object required for the escape velocity to equal the speed of light. This radius defines the event horizon of a black hole. Within the Schwartzchild radius, i.e., the event horizon, nothing can escape, not even light. This area will appear dark, hence the term black hole.

singularity—A point of zero dimensions and infinite density. The center of a black hole.

solar system—A star-centered system containing planets and planetary remnants. The solar system in which the sun is centered contains the planets Mercury, Venus, Earth, Mars, Jupiter, Saturn, Uranus, Neptune and Pluto, and the asteroid belt (between Mars and Jupiter) and a cloud of comets surrounding the system referred to as the Oort cloud.

spacetime—The union of three spatial dimensions and one temporal (time) dimension composing the fabric of the universe. Also referred to the spacetime continuum of the universe,

steady state theory—Theory that the universe is infinitely old and maintains a constant density throughout by the creation of matter in the vacuum of space, as the universe expands. This theory is somewhat obsolete.

strong (nuclear) force—One of the four fundamental forces of nature, it's responsible for bonding together atomic nuclei.

supernova—A star that explodes. The blast results from a star's rapid transformation into a neutron star. The Latin root *nova* means "new star."

superstring theory—Theory that the smallest constituents of matter are not point-like particles but tiny vibrating filaments of strings. A particle's mass and charge depend on the resonant frequency of the oscillating string. The strings are theorized to be curled-up higher dimensions of spacetime.

tritium—An isotope of hydrogen containing two neutrons.

weak (nuclear) force—One of the four fundamental forces of nature, it is responsible for beta decay, where a neutron will spontaneously transform into a proton, emitting an electron and a boson in the process.

white dwarf—The stable, end result of an average-mass star following its red giant phase. A white dwarf results from the stellar core being exposed after the outer layers have been blown off. White dwarfs are typically the size of the earth. The Pauli exclusion principle prevents further collapse of a white dwarf star by preventing the electrons from moving any closer together.

WIMPs (or Weakly Interacting Massive Particle)—Theoretical form of dark matter in the form of subatomic particles, such as neutrinos, that proliferate the universe and (may) contain mass. If neutrinos contain even the slightest amount of mass, they may account for most of the mass of the universe. There are approximately a half billion neutrinos for every proton and neutron within the universe. WIMPs are "weakly interacting" because they very seldom react with ordinary matter and "massive," not because they are huge and heavy, but because they may contain the slightest amount of mass.

APPENDIX A

Hubble Constant versus Age of the Universe

The Hubble law may be used to determine the age of the universe; however, an accurate measure of the Hubble constant must be determined, since the Hubble constant is indirectly proportional to the age of the universe:

$$v = H_0 d$$

where H_0 is the Hubble constant in units of $km \cdot s^{-1} \cdot Mpc^{-1}$ and a megaparsec (Mpc) is equivalent to $3.26 \cdot 10^6$ light-years, and the speed of light (c) is equivalent to $3.00 \cdot 10^8$ m/s. Convert Mpc to km:

$$pc = 3.26 ly$$

$$c = 3.00 \cdot 10^8 \frac{m}{s} \cdot \frac{km}{1000m} = 3.00 \cdot 10^5 \frac{km}{s}$$

$$v = \frac{d}{t} \Rightarrow c = \frac{d}{t}$$

$$d = ct$$

where t = 1 yr for a distance (d) of 1 light-year

$$d = ct$$

$$d(1ly) = 3.00 \cdot 10^5 \frac{km}{s} \cdot 1yr \cdot 3600 \frac{s}{hr} \cdot 24 \frac{hr}{d} \cdot 365 \frac{d}{yr}$$

$$d(1ly) = 9.46 \cdot 10^{12} km$$

$$\therefore d(1pc) = 3.26(9.46 \cdot 10^{12})km = 3.08 \cdot 10^{13} km$$

$$d(1Mpc) = 3.08 \cdot 10^{13} \cdot 10^6 = 3.08 \cdot 10^{19} km$$

$$\therefore H_0 \equiv \frac{km}{s \cdot Mpc} = \frac{km}{s(3.08 \cdot 10^{19})km} = 3.24 \cdot 10^{-20} \frac{1}{s}$$

or

$$t = \frac{1}{H_0} \equiv 3.08 \cdot 10^{19} s \cdot \frac{hr}{3600s} \cdot \frac{d}{24hr} \cdot \frac{yr}{365d}$$

$$t = \frac{1}{H_0} \equiv 9.78 \cdot 10^{11} yr$$

$$\therefore H_0 = 70 \frac{km}{s \cdot Mpc} \Rightarrow t = \frac{9.78 \cdot 10^{11}}{70} yr$$

$$= 1.40 \cdot 10^{10} yr$$

$$= 14.0 \cdot 10^9 yr$$

So, a Hubble constant of $H_0 = 70$ km·s^{-1}·Mpc^{-1} puts the age of the universe at t = 14.0 billion years.

APPENDIX B

Blackbody Spectrum and the Planck Radiation Formula

The energy distribution detected by the COBE satellite agrees with quantum theory for blackbody radiation. The measured curve follows the Planck radiation formula:

$$E(f) = \frac{8\pi f^2}{c^3} \cdot \frac{hf}{e^{\frac{hf}{kT}} - 1}$$

where
$\pi = 3.14159$
$c = 2.99810^8$ ms^{-1} (speed of light in a vacuum)
$h = 6.62610^{-34}$ Js (Planck's constant)
$k = 1.38110^{-23}$ JK^{-1} (Boltzmann's constant)
$T = 2.73$ K (average temperature of the universe)

REFERENCES

Adams, Fred and Laughlin, Greg. *The Five Ages of the Universe: Inside the Physics of Eternity*. New York: The Free Press, 1999.

Breithaupt, Jim. *Cosmology*. Chicago: Contemporary Publishing, 1999.

Chang, Raymond. *Chemistry, 3rd ed*. New York: Random House, 1988.

Ferris, Timothy. *Coming of Age in the Milky Way*. New York: William Morrow and Company, 1988.

Ferris, Timothy. *Galaxies*. New York: Stewart, Tabori & Chang, 1982.

Ferris, Timothy. *The Whole Shebang: A State-of-the-Universe(s) Report*. New York: Simon & Schuster, 1997.

Filkin, David. *Stephen Hawking's Universe: The Cosmos Explained*. New York: BasicBooks, 1997.

Gell-Mann, Murray. *The Quark and the Jaguar*. New York: Freeman, 1994.

Goldsmith Donald. *The Astronomers*. New York: St. Martin's Press, 1991.

Greene, Brian R. *The Elegant Universe: Superstrings, Hidden Dimensions, and the Quest for the Ultimate Theory*. New York: W. W. Norton & Company, 1999.

Gribbin, John. *The Case of the Missing Neutrinos and Other Curious Phenomena of the Universe*. New York: Fromm International, 1998.

Guth, Alan H. *The Inflationary Universe*. Reading, Mass.: Perseus Books, 1997.

Hawking, Stephen. *A Brief History of Time*. New York: Bantam Books, 1988.

Hawking, Stephen. *The Theory of Everything*. Beverly Hills: New Millennium Press, 2002.

Hawking, Stephen. *The Universe in a Nutshell.* New York: Bantam Books, 2001.

Lerner, Eric J. *The Big Bang Never Happened.* New York: Vintage Books, 1991.

Menzel, Donald H. and Pasachoff, Jay M. *A Field Guide to Stars and Planets.* Boston: Houghton Mifflin Company, 1983.

Motz, Lloyd and Nathanson, Carol. *The Constellations: An Enthusiast's Guide to the Night Sky.* New York: Doubleday, 1988.

Rees, Martin. *Just Six Numbers: The Deep Forces that Shape the Universe.* New York: BasicBooks, 2000.

Riordan, Michael and Schramm, David N. *The Shadows of Creation: Dark Matter and the Structure of the Universe.* New York: Freeman, 1991.

Sagan, Carl. *Cosmos.* New York: Ballantine Books, 1980.

Sagan, Carl. *Pale Blue Dot: A Vision of the Human Future in Space.* New York: Random House, 1994.

Schmidt, Frank W., Henderson, Robert E. and Wolgemuth Carl H. *Introduction to Thermal Sciences: Thermodynamics/Fluid Dynamics/Heat Transfer.* New York: Wiley, 1984.

Serway, Raymond A. *Physics for Scientists and Engineers, 2nd ed.* Philadelphia: Saunders College Publishing, 1986.

Weinberg, Steven. *The First Three Minutes.* New York: BasicBooks, 1988.

Index

0-595-30125-8

www.ingramcontent.com/pod-product-compliance
Lightning Source LLC
Chambersburg PA
CBHW030839180526
45163CB00004B/1385